Conservation and Restoration of Built Heritage

Built Heritage and Geotechnics
Series editor: Renato Lancellotta

The Restoration of Ghirlandina Tower in Modena and the Assessment of Soil–Structure Interaction by Means of Dynamic Identification Techniques
Rosella Cadignani, Renato Lancellotta and Donato Sabia

Correction of Differential Settlements in Mexico City's Metropolitan Cathedral and Sagrario Church
Efraín Ovando-Shelley and Enrique Santoyo

The Tower of Pisa
History, Construction and Geotechnical Stabilization
J.B. Burland, M.B. Jamiolkowski, N. Squeglia and C. Viggiani

The Cathedrals of Pisa, Siena and Florence
A Thorough Inspection of the Medieval Construction Techniques
Pietro Matracchi and Luca Giorgi

Conservation and Restoration of Built Heritage
A History of Conservation Culture and Its More Recent Developments
Salvatore D'Agostino

For more information about this series, please visit: www.routledge.com/Built-Heritage-and-Geotechnics/book-series/BHG

Conservation and Restoration of Built Heritage

A History of Conservation Culture and Its More Recent Developments

Salvatore D'Agostino

Routledge
Taylor & Francis Group
LONDON AND NEW YORK

Cover image: Notre-Dame De Paris Summer France Stock Photo (Edit Now) 1376487728

CRC Press/Balkema is an imprint of the Taylor & Francis Group, an informa business

Library of Congress Cataloging-in-Publication Data
Names: D'Agostino, Salvatore, author.
Title: Conservation and restoration of built heritage: a history of conservation culture and its more recent developments/Salvatore D'Agostino.
Description: Boca Raton: CRC Press, [2021] | Series: Built heritage and geotechnics | Includes bibliographical references and index.
Identifiers: LCCN 2021021499 Subjects: LCSH: Historic buildings–Conservation and restoration.
Classification: LCC TH3411 .D24 2021 | DDC 363.6/9–dc23
LC record available at https://lccn.loc.gov/2021021499

Published by: CRC Press/Balkema
Schipholweg 107C, 2316 XC Leiden, The Netherlands
e-mail: Pub.NL@taylorandfrancis.com
www.crcpress.com
www.taylorandfrancis.com

ISBN: 9780367750954 (Hbk)
ISBN: 9780367750961 (Pbk)
ISBN: 9781003160960 (eBook)

DOI: 10.1201/9781003160960

Typeset in Goudy
by Deanta Global Publishing Services, Chennai, India

To my wonderful family

Contents

Foreword

This is the fifth in a series of volumes on Built Heritage and Geotechnics, intended to reach a wide audience: professionals and academics in the fields of civil engineering, architecture and cultural resources management, and particularly those involved in art history, the history of architecture, geotechnical engineering, structural engineering, archaeology, restoration and cultural heritage management – and even the wider public.

The motivations behind this series rely on the fact that preservation of built heritage is one of the most challenging problems facing modern civilisation. It involves, in inextricable patterns, various cultural, humanistic, social, technical and economic aspects. The complexity of the topic is such that a shared framework of reference is still lacking among art historians, architects and structural and geotechnical engineers.

This is proved by the fact that, although there are exemplary cases of integral preservation of structural components with their static and architectural function, as material testimony to the knowledge, culture and constructional techniques of the original historical period, there are still examples of uncritical confidence in modern technology, which has led to the replacement of previous structures with new ones, which only preserve the iconic appearance of the original monument.

For these reasons, publishing short books on specialised topics, such as well-documented case studies of restoration or detailed overviews of construction techniques, designed to provide material evidence of our knowledge of the historical period in which the monuments were built or of the evolution of the cultural approach of conservation, may be highly significant.

In this context, this volume, authored by Salvatore D'Agostino with a preface by Giovanni Calabresi, provides a critical reading of the history of conservation from antiquity to its more recent achievements, including the strategies adopted by leading European countries in this field.

In particular, the volume highlights how, during the second half of the last century, the preservation process also involved engineering, the science of making practical application of knowledge, which, for a long time, made an uncritical use of techniques and materials and devised interventions on historical heritage that were heavily invasive.

In this respect, examples of damage caused by the improper use of new materials, and specifically reinforced concrete, are underlined, showing how it was necessary to wait until the last decades of the previous century to see engineering promote a process of critical revision which, gradually and not without difficulty, merged into the modern theory of conservation.

The volume also deals with a particular aspect that deserves special attention: the problem related to seismic risk to which Italy, Greece and Portugal are particularly prone.

Problems that emerge during the crisis and reconstruction phases are dealt with in detail, as well as the need for properly scheduled maintenance, since this latter approach always constitutes an improvement in the performance of the monument and is the most appropriate tool for the conservation of built heritage.

Lastly, the volume provides a series of case studies of restoration of many outstanding monuments.

The book originates from a lengthy research project carried out by the author, and his immense experience in the field is reflected in his ability to explain complex concepts in a clear and effective way. For this reason, this book is addressed not only to professionals and academics in the wider fields of civil engineering (both geotechnical and structural engineering), architecture, art history, the history of architecture, restoration and cultural heritage management, but it may also give a solid background to undergraduates and graduate students involved in the preservation of monuments.

Renato Lancellotta
Series editor

Preface

The conservation of historical and monumental buildings is a topic where the contribution of engineering is of fundamental importance. In this context, geotechnical and structural engineers cannot operate independently, as has also been illustrated in the famous Rankine Lecture by Jean Kerisel in 1974, which aroused great interest among the geotechnical community and a few years later led the International Society of Soil Mechanics and Geotechnical Engineering (ISSMGE) to set up the Technical Committee on the Preservation of Monuments and Historic Sites. However, it should be noted that the contribution of engineers is generally viewed (and considered by the institutions responsible for safeguarding built heritage) from an exclusively technological perspective. The progress made over recent decades in the fields of construction materials and operational technologies in structural and geotechnical engineering has contributed to the conservation of historic buildings by applying reinforcement solutions that fail to respect traditional construction techniques and the original design concept of the structures. This approach to the consolidation of historic and monumental buildings weakened by their age does not respect the principle of preserving the memory of the building as material evidence of its era, which becomes irreparably lost, and is therefore contrary to the current approach to conservation, as well as the vision of Jean Kerisel, whose immense humanism and passionate love of the art of building in the past comes across very clearly in his beautiful books.

Furthermore, experience has shown that the addition of new materials often loses its efficacy over time and can even cause permanent damage to the works it was supposed to save.

In the geotechnical field especially, this type of intervention has gradually become more widespread over the last 60 years due to the development of new technologies. Micropiles (also known as root piles), which can be made through existing masonry structures, have been widely used to consolidate historic buildings damaged by the deformation of the soil or for the protection of structures generally considered to be at risk; the original shallow foundations are replaced by deep ones, which are not affected by the behaviour of the upper layers of the foundation soil, while the application of concentrated forces significantly alters the stress state of the overlying structure.

There are cases where interventions of this type are necessary to save a monumental building, as was the case for the Ponte Vecchio in Florence, when the Arno riverbed was deepened following the tragic flood of 1966. However, in many other situations, micropile underpinning is used without any real need, irreparably altering the historic building, just because it is a far simpler solution and, paradoxically, less expensive than looking for the cause of soil deformation and studying whether it could be resolved. In defence of this choice, it is often stated that the intervention is invisible; a purely visual conception of conservation seems to prevail. Among the interventions carried out in the past on important monumental works, a conservation approach that respects the original construction features, as in the case of the Tower of Pisa, the Torre Ghirlandina in Modena and the Ponte Milvio in Rome, is rarely applied.

For almost two centuries, the conceptual assumptions underlying heritage conservation have been the subject of debate, modification, theoretical formulations and directives in conferences, institutions and international organisations, especially in Europe. It is therefore to be hoped that geotechnical engineers should be involved in, and take an active part in, the consolidation and conservation of historic building projects, not only offering passive support with their knowledge of advanced technologies.

This book examines the contribution of engineering in general and particularly of structural engineering to conservation issues, which is why it has been included in a series devoted to geotechnical problems, promoted by the Technical Committee on Preservation of Monuments and Historic Sites of ISSMGE. Geotechnical and structural engineering are inseparable elements of civil engineering, closely interconnected in seeking solutions to the frequently difficult problems raised by the preservation of the historical heritage, to which the author of the book, Salvatore D'Agostino, has dedicated most of his academic and professional life with great passion as Professor of Structural Engineering at the historic Federico II University of Naples. The book provides ample evidence of his experience and knowledge of cultural trends, conservation theories and guidelines in Italy and many other European countries. The section devoted to Italy undeniably occupies the lion's share, not just because this represents the bulk of his work, but also and above all because only Italy still preserves so much material evidence of the numerous civilisations that have emerged in its territory over a period of almost three millennia, and whose conservation is of crucial importance.

D'Agostino has a deep-rooted knowledge of traditional construction techniques and their evolution over the centuries and is passionately concerned about their conservation. It is in this spirit that, fifteen years ago, he founded the Italian Association for the history of engineering, of which he is president. I believe that reading this book can open up new horizons of knowledge in the field of the conservation of historic buildings.

Giovanni Calabresi

Biography

Salvatore D'Agostino has taught Structural Analysis at Federico II University in Naples for about 40 years. He has also been involved in building restoration, and in 1992 founded the Centro Interdipartimentale di Ingegneria per i Beni Culturali (Interdepartmental Centre for Engineering for Cultural Heritage) at the same university. He founded, and is chairman of, the Associazione Italiana di Storia dell'Ingegneria (Italian Society of the History of Engineering). He edited the first text on engineering for cultural heritage, published by Edizioni del Mulino in 2018. He is author of about 170 academic publications.

Acknowledgements

The author would like to express his heartfelt gratitude to Professor Giovanni Calabresi, an expert in geotechnical engineering, for his invaluable and generous assistance with the preparation and drafting of this volume. Professor Calabresi has displayed the rigour of a consummate professional and the kindness of a friend.

With all my thanks,
Salvatore

Translation
Colum Fordham

The conservation of monumental architecture in Europe

Introduction

The conservation of built heritage is a fundamental need of the western world. The whole of Europe is dotted with civic and religious monuments and has a connective tissue made up of hundreds of historic towns and cities and thousands of villages. Europe has the largest number of UNESCO World Heritage sites due to their immense historic importance and the beauty of the landscape.

The concept of conservation is intrinsically complex and has evolved over time, because it is directly conditioned by the vision of history in various epochs. Humans have always aspired to construct monuments that they would like to be eternal, but instead endure the violence of time and man-made destruction.

The following is a brief summary of the history of conservation within the context of western culture over different epochs.

Conservation in the ancient world developed as part of a unitary concept going back thousands of years and, in the western hemisphere, it was fundamentally based on the Roman architectural tradition. Despite the infinite number of local adaptations, this tradition evolved without major innovations for about two millennia and only in the late 19th century did it undergo a radical change, both in terms of the idea of construction and in terms of materials and techniques. This led to a new technical and scientific culture of building which made old concepts, and traditional materials and techniques obsolete.

As a result of the compartmentalisation of knowledge, conservation has always veered towards the art historical sphere while the development of the mechanics of construction, industrial materials, steel, concrete and the engineering of technological plants, with the emergence of rational architecture, has gradually modified the concept of inhabiting space. Despite continuing to play key functions at a cultural and social level, built heritage has become a part of archaeological heritage due to the architectural concept.

Nevertheless, since material history has been recognised as an essential component of the evolution of civilisation, conservation and restoration have acquired new cultural significance which, with the emergence of archaeometry, have led to a new and fertile relationship between scientific research and conservation. During the second half of the twentieth century, this process has also involved engineering, the science of doing and the possible, which has long used, in an uncritical manner, techniques and materials that conflict entirely with the history of conservation, leading to the use of highly invasive interventions on historical heritage. Only in the last few decades of the twentieth century has engineering begun a process of critical reassessment which has gradually led, not without problems, to the emergence of engineering geared towards cultural heritage, in line with Conservation Theory (D'Agostino 2017).

The time is therefore ripe for a brief critical overview of the history of conservation and its most recent developments so that engineering can fully collaborate with the art historical culture of the conservation of Europe's vast historical built heritage.

This volume will analyse the modern hypothesis of history as the science of humanity, as well as the recent development of material history as the promoter of archaeometric research.

The volume will also illustrate how conservation is currently a cornerstone of European culture and the strategies to preserve it, highlighting the problems linked to seismic risk to which Italy, Greece and Portugal are particularly susceptible.

From this perspective, structural and geotechnical engineering have both become essential reference points for the conservation of material history.

Their evolution over the twentieth century will be analysed and the cultural imbalance caused by reinforcement measures carried out during this century will be illustrated. An analysis will be presented of the damage caused by the improper use of new materials, especially of reinforced concrete. Through a detailed examination of seismic risk, the various problems that come to light during the phase of emergency and reconstruction will be identified. Planned maintenance, which always leads to an "improvement" of the monument, will be identified as the most suitable solution to the conservation of built heritage. The different relationship between conservation and safety will be emphasised by comparing the situation in several countries. Lastly, several case studies will highlight the cultural and scientific requirements involved in the conservation of built heritage.

Chapter 1

A brief history of conservation

1.1. The conservation of antiquity

It may appear contradictory, but humankind is increasing its knowledge of space more than that of the earth, both in terms of its geological depth and in terms of the transformation and humanisation of its species. Until several decades ago, the Sumerian ziqqurats, which are dated to around 5000 BC, were considered to be the oldest evidence of architecture. The discovery of the site of Göbekli Tepehas pushed this date back to about 9500 BC. The purpose of this archaeological site, situated in Turkey, still remains uncertain. It contains stone circles made of stelae or pillars 3 metres high, each weighing about 15 tons. The stelae are carved with animals and geometric symbols. They probably represent individuals, while the stone circles represent assemblies (Figure 1.1). Research has indicated the presence of numerous other stelae. Firstly, this discovery shows the state of material culture achieved by humans through quarrying stone, finishing the blocks, and raising them, and the sophisticated level of sculpture. This suggests the existence of even older works and settlements.

The geology of the site suggests that it was intentionally buried in about 8000 BC due to abandonment or due to military or religious events. However, the burial may reflect the intention to preserve the site so that its memory could be conserved for the future. It is the earliest known example of conservation. The Bible also contains examples of conservation and, obviously, of handing down the memory of the past. The collection of the 12 stones from the river Jordan after the waters had been parted also represents a desire to commemorate an extraordinary event. Obviously, there are also examples of "damnatio memoriae" in the Bible. In antiquity, the link between sacred and profane was so close that their boundaries became blurred, just as religion and rights were not distinctly separate.

The whole of Western history is marked by the lengthy stratification of monuments, preserved with the aid of restoration, which is the material tool by which conservation is carried out. Restoration originated from the need to repair tools and was obviously used to hand down cult objects of immense symbolic significance. The first repairs date back to 7000–8000 BC, while ancient Greek vases were restored using techniques that were far more elaborate than those used for everyday objects. Similarly, metal clamps and dowels were inserted to repair cracks in marble. The picture that emerges is of an ancient world which, in many ways, displays similarities with the modern practice of restoration, contradicting the theory, put forward for example by Viollet le Duc, that restoration originated in the nineteenth century (Pergoli Campanelli 2015).

Before the Romans, the Greeks were aware of living in a country strewn with the ruins of partially concealed cities and tombs, both intact or already violated, of previous inhabitants.

DOI: 10.1201/9781003160960-1

Figure 1.1 The stelae of Göbekli Tepe – Wikipedia

Many of the objects – often mutilated and extremely ancient – from these discoveries can be identified in the inventories of sanctuaries. More frequently, they were part of the booty seized from Athens and the rest of Greece by the Persians which, following the campaigns of Alexander the Great, were returned to their places of origin and restored. Restoration is the material expression of a conservative approach, and the periodic maintenance of buildings was already regarded as extremely important in antiquity, as shown by the restoration work of the Sanctuary at Delos (A. Melucco Vaccaro 1989).

The sacred sphere constituted one of the main incentives for conservation in antiquity. By establishing rules and rituals, liturgy also gave rise to the preservation of objects of worship, first and foremost simulacra, as is emphasised by Pausanias, who, writing in the second century AD, mentions ancient wooden statues that had survived as a result of careful maintenance.

In the Roman world, the emperors issued numerous decrees related to the protection and conservation of monuments. The passage by Pergoli Campanelli, already cited above, analyses Roman law in relation to the many links with safeguarding both private and public monuments. Cicero mentions specific lists of works of art for state heritage that included war booty, while the office of "curatores rei pubblicae" – responsible for the care and rebuilding of derelict structures – can be dated to the first century AD. The point of contact between Roman law and Christian culture began to develop between the fourth and fifth centuries AD. It is worth emphasising that Constantine's edict was issued in AD 313. "This was the period when there was synergy between ancient Roman law and Christian culture which was to be of crucial importance to the formation of many concepts underlying modern restoration theory" (Pergoli Campanelli, op. cit. 2015). A famous example is the emperor Julius Valerius Maiorianus (420–461), who called on Roman citizens to preserve ancient buildings using all possible means. For the construction of the Basilica of St Peter, Constantine himself was forced to undertake the respectful burial of monumental cemeteries, and the conservation of St Peter's tomb is emblematic. At the same time, a cult of relics evolved which, in its quest for authenticity, laid the foundations of Western culture in terms of the importance of and respect for materials. It was undoubtedly a fluctuating history where enlightened actions alternated with extremely destructive and disrespectful episodes with regard to works of art. However, on careful reflection, "damnatio memoriae" is a human reaction designed to exorcise the preservation of memory of the reviled enemy. The destruction of the Domus Aurea out of hatred for Nero is a classic example. The history of ancient architecture is marked by destruction, plundering and drastic alteration, even though there are exceptions, such as the Temple of Hera Lacinia at Croton, which may be due to religious reasons rather than attempts at conservation. This approach was the result of the sense of history among the ancients. There was primarily a hagiographic vision closely linked to myths and rituals to be illustrated in this specific civic and religious climate.

The Ostrogothic kings displayed great sensitivity to conservation which benefited from the presence of Flavius Magnus Aurelius Cassiodorus Senator, who operated initially at the court, in continuity with Roman law, and subsequently in the Christian world of the monastery that he himself founded. His writings displayed considerable awareness of the theory of restoration with regard to ancient monuments and new works; his emblematic maxim was "only the newness of the buildings should distance them from the works of the ancients" (*nova construere, sidamplius vetusta servare*). There has been much controversy among historians who have different verdicts on the policy pursued by Pope Gregory I, known as Pope Gregory the Great. He was long considered, following John of Salisbury (1120–1180), to be a destroyer of monuments, a view that was refuted by Platania (1421–1481) and Pompeo Ugonio (1587–1613) until the eighteenth century with the writings of Girolamo Tirabocchi. According to Pergoli Campanelli (op. cit. 2015), the theory that the pope espoused conservation appears to be backed up in the correspondence of Pope Gregory and, in particular, in a letter written to Abbot Militto and a letter to Serenus, Bishop of Marseilles. Campanelli concludes: "Through his enlightened measures he encouraged the transformation and reuse of many ancient monuments otherwise destined to suffer a deplorable fate".

The difference between East and West became even starker: in Constantinople, the construction of new buildings led to the destruction of a large part of its ancient architecture, whereas in Rome there was already concern about restoring the monuments of a glorious and extraordinary past. As already mentioned, Rome had experienced a chequered history where enlightened actions went hand-in-hand with extremely destructive measures. However, the seeds had been sown, and the history of Western restoration would take a different course from events in the eastern Roman Empire.

Lastly, the discoveryof "material" conservation had enabled, ever since antiquity, the rules of the art to be handed down from generation to generation. The need to carry out repairs initially and only later to undertake restoration underlined the unstinting efforts of humans to assert their own existence and history, and prevent the constant and inevitable deterioration of material in order to convey the image of the work. Ever since prehistory, conservation has concerned the material nature of the work and its image which is imbued with historical significance. Another aspect of conservation is reuse, which is not restricted to economic factors. In antiquity, historic works were often "tampered with" and "transformed" by adaptive reuse. This was due not only to issues of functionality but also to political requirements and to matters of taste. However, there are also examples of reconstruction without modifications, such as the case of the Temple of Apollo at Pompeii, destroyed by the earthquake in AD 62.

The Basilica Emilia of 179 BC has survived in the wake of renovation work carried out from the era of Augustus to Tiberius and later following the sack of Rome by Alaric in AD 410. These are just a few examples of a widespread and amply documented practice that continued over the centuries, as is shown by the basilica of S. Salvatore in Spoleto (Italy), which was repeatedly restored up to the time of Lombard rule in the eighth century AD.

Another facet of cultural sharing in antiquity is the huge Roman production of copies of Greek statues, bearing in mind that a copy had an intrinsic value and was not merely a reproduction, as has been argued by proponents of historicism who underline the unrepeatable nature of art works; this is one of the many examples of cultural discontinuity between past and present, although it has been brought back into question by the spread of the digital aesthetic. One final aspect of conservation lies in cultural continuity, which emerges in artistic and celebratory events. The Roman triumphal arch is a shining example (Figure 1.2). It played a highly significant and symbolic role in passing down architectural forms from antiquity to the modern era. Its importance also extended to the figurative arts (Figure 1.3) and even to the ephemeral constructions of triumphal festivals.

Figure 1.2 Rome, the Arch of Constantine – Wikipedia

Figure 1.3 Padua, Ovetari Chapel, "The Miracle of St. James" by Andrea Mantegna – Wikipedia

The Roman triumphal or honorary arch developed from the Republican *fornices* and became an architectural feature of the whole empire. with famous examples such as the Arch of Trajan at Benevento and the Arch of Constantine at Rome. This would be prolonged infinitely with the topos of the image of Western culture represented by the Triumphal Procession, which led to the ancient ceremony being re-enacted during the Humanist Age and the Renaissance. Leading figures in this process included architects such as Alberti, Palladio and Serlio, whose treatise written in 1540 contained the first systematic series of triumphal arches in Rome and Italy (Figure 1.4). Further treatises have been written on the subject by authors such as Milizia and Diderot, down to the present (De Maria 1988). Triumphal arches were not confined to Europe and have been built all over the world. The Roman honorific arch, like the triumphal arch, is a monument that bears an infinite number of historical, emblematic and figurative references, ensuring its enduring popularity, which has no parallel in the history of architecture. Another type of monument that enjoyed similar popularity, despite being more circumscribed and anthropologically different, was the Roman basilica, the model and forerunner of Early Christian churches and much of subsequent Western sacred architecture.

In conclusion, conservation was a response, even in antiquity, to a need perceived by civilisation, a need for historical continuity and a symbol of civilisation. It took on many different guises, such as copying, reuse and imitation, and was always associated with restoration which, as already mentioned, together with maintenance – also widespread and admired in antiquity – represent the material form of conservation.

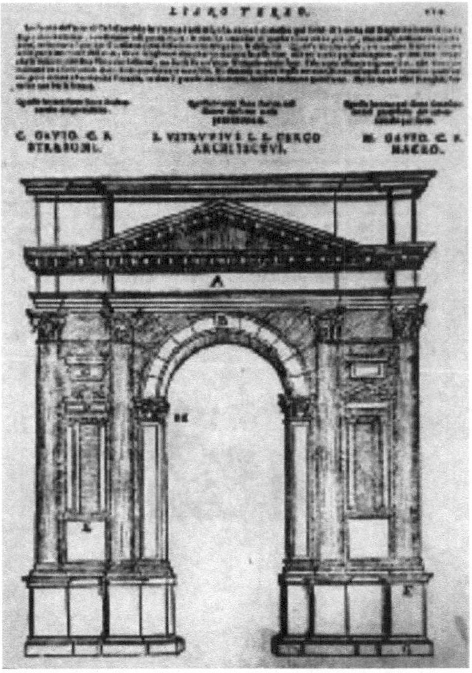

Figure 1.4 The Triumphal Arch in Palladio's treatise – Palladio; The Four Books of Architecture

1.2. Conservation from the Middle Ages to the modern era

A distinctive feature of the Middle Ages was the reuse of elements or parts of ancient structures, mainly dating to the Roman period, and decorative elements, termed *spolia*, which were rearranged in other buildings. Reuse is widely found in all regions of Europe from Gallia to Asia Minor, utilising spolia in newly constructed buildings or to modify and adorn existing buildings. "The Latin term 'restaurare' was used in the Middle Ages, like the terms 'instaurare', 'renovare', 'reficere' and 'adoptare', which could be generally translated as adjustment, without implying a return to the building's original identity and to the documentary and aesthetic uniqueness of the ancient monument" (Cordaro 2000). This is also evident due both to the medieval concept of the artist as "homo faber", not yet endowed with creative autonomy, and to the fact that the notion of enhancing material in historical and scientific terms was still a remote hypothesis. There were compelling reasons for reusing spolia, which were recognised as possessing an intrinsic value that should be recycled and preserved. The lack of a concept of the "uniqueness" of the artistic artefact encouraged reuse and reworking, as in the case of the Roman sarcophagus in the Duomo in Salerno. It is hard to establish whether the wealth of spolia and building material, especially bricks, contributed to the decadence of artistic practice and the limited construction of arches, or vice versa. A classic example is the work of Desiderio di Montecassino (1058–1086), who purchased columns, bases and precious coloured marble in Rome for the reconstruction of the church of San Benedetto, which featured a profusion of ornaments and lavish decoration for which he commissioned highly skilled artists and artisans from Constantinople. Desiderio was heavily influenced by the precious nature of the artefacts but also by their significance as akin to relics of antiquity, thus undertaking work that could be viewed as conservation. Similar stories could be told of the Abbey of Saint Denis or Canterbury Cathedral, reflecting the general cultural ethos of medieval Europe. Indeed, the exemplary edict of the Senate of Rome of 1162 expressly forbade people from demolishing or inflicting any kind of damage on Trajan's column (Figure 1.5), a sign that it represented the very notion of *romanità* (the Roman spirit). In Italy, with the re-emergence of flourishing city-states, the display of ancient walls and gates took on the character of intentional conservative restoration through their adornment with ancient artefacts, which were sometimes re-worked. This is especially evident in city gates such as Porta Marzia at Perugia, but also at Milan, Pavia, etc.

A radical change took place with the emergence of humanism through the concept of humans and their dignity as authors of their own destinies, a constant and crucial reference point of ideological reflection. Philological humanism, geared towards recovering, studying and publishing classical texts, manifested itself as the "preservation of literature" and would pave the way for the advent of modern humanism when, in the mid-fifteenth century, ancient literary ideals would be transferred to cultures in the vernacular. Petrarch (Francesco Petrarca) rediscovered ancient codices, tried to re-establish Latinity as a linguistic instrument and displayed great interest in ancient buildings, celebratory monuments and equestrian statues, sensing the allure of ruins. His letters reflect the philosophical love of knowledge and understanding of antiquity. In a letter written in 1468 to the Doge of Venice, in which he donated his library (482 volumes in Greek and 264 in Latin) to the city, Cardinal Bessarion wrote "So great is their strength, their dignity, their majesty and

Figure 1.5 Rome, Trajan's Column – Wikipedia

ultimately their sacredness, that if there were no books, we would all be coarse and ignorant with no memory of the past, without any example on which to rest. We would have no knowledge of human and divine things".

This period witnessed the emergence of philology, which tended to make a clear distinction between different eras, distancing the present from the past. At the same time, the work of art became the manifestation of an ideal that aspired to formal perfection. Works of historic significance were created that reflected the career paths of individual artists, such as Lorenzo Ghiberti's *The Commentaries* and, subsequently, Giorgio Vasari's *Lives*. Artists studied the works of the past to broaden their knowledge and grasp the works' compositional and constructional principles. Thus, while destruction and looting went on unabated, Renaissance architects celebrated knowledge of the past with a large corpus of drawings and designs, a practice that would continue over the following centuries. They preferred to work on the unfettered completion of an ancient work of architecture rather than its philological conservation, since modern concepts of authenticity and inviolability had not yet evolved. Examples include Leon Battista Alberti's work on the Tempio Malatestiano and the church of Santa Maria Novella in Florence, as well as the completion of the ruins of the Theatre of Marcellus attributed to Peruzzi.

Raphael was appointed prefect of antiquities in Rome by Leo X in 1515. He displayed a philological approach to the restoration of parts of ancient buildings that had been damaged or destroyed to revive their ancient splendour, if for no other reason than to be part of the virtual reality of the volume of drawings presented to the pope. The aim of completing a work can be better appreciated by considering the relationship between painting and sculpture during the Renaissance. Vasari's assessment is a case in point: he expressed his appreciation for the "restored antiquities" (*anticaglie restaurate*) following the arrangement of the courtyard of Palazzo della Valle, whose façade was adorned with statues, cornices, reliefs and the front panels of sarcophagi, as at Villa Medici al Pincio (1576) and Villa Doria Panphili. The statues were discovered in a fragmentary state, with irregularities, missing parts and a patina caused by weathering. They underwent drastic cleaning, and arms, heads and legs found in excavations or commissioned from artists were added. The reconstructions of the statue group *Laocoön and His Sons* and the legs of the *Farnese Hercules* are particularly famous. The restoration was treated as an exercise in completion, and the work was entrusted to an artist. This approach would continue until the eighteenth century, when concepts of authenticity and uniqueness began to gain ground. The Counter-Reformation and the Baroque were marked by yet another approach. After the Council of Trent in 1545, the Catholic Church had to reassert its own traditions through a transformation of meanings. Restoration was conceived as an attempt to safeguard the memory of ancient Christianity and as an essential means of reaffirming its values. In 1588, Sixtus V ordered the demolition of the Septizodium since it was a symbol of paganism; restoration was carried out on Trajan's column, and a statue of St Peter was placed at the very top. Simultaneously, an exemplary papal bull was issued by Sixtus V in 1587 which used the term "conservation" to safeguard the pinetum at Ravenna, while the Chirograph (personal writ) issued by Innocent X in 1647 sanctioned the conservation of and respect for works built by Early Christian emperors. A complex relationship also emerged between ancient and modern in the restoration work carried out by architects when asked to intervene on ancient monuments. An illustrative case is the complex intervention – unfortunately incomplete – carried out by Borromini on the Basilica of San Giovanni in Laterano to celebrate the Jubilee of 1650. Although he did not intervene in the transept, which had its own unity, the artist invented spatial solutions which fully conform to the precepts of Baroque culture. The historic building was still an "evolving" structure before it attained its present-day appearance as an "archaeological construct" following the revolution in construction techniques of the nineteenth and twentieth centuries.

1.3. Conservation and restoration from the eighteenth to the twenty-first century

The first leading figure in the modern era was Pietro Edwards (1744–1821), a painter born in Italy but of English origin, who worked extensively in Venice. He was a champion of an extremely modern approach to restoration and pioneered the theory of minimal intervention together with great respect for the work of art, using materials and techniques that conform to the original. A member of the Accademia di San Luca and director of the Restoration of the Public Pictures at Venice, as well as the founder of a collection that would lead to the present-day Gallerie dell'Accademia, he was the author of a document that set out the obligations of restorers, which can be rightfully considered the forerunner of the modern Italian Charters for Restoration (*Carte del Restauro*). Obviously, this modern approach did not concern historic architecture, but it does reflect cultural maturity, which was particularly respectful of the material substance of the work of art.

In the early nineteenth century, France played a leading role in the reconstruction and restoration of the ruins left by the revolution. However, arbitrary interventions tended to prevail. The Basilica of Saint Denis was seriously damaged and François Debret, who worked on the building from 1813 to 1846, carried out renovation work and made additions that seemed largely arbitrary. He also reconstructed the north tower, which was unstable and was demolished in 1846. In the church of Saint Ouen, Grégoire regarded the sixteenth-century towers as inauthentic and demolished them, completely redesigning the façade. These judgements led Victor Hugo to make the following harsh pronouncement in 1825:

> We have to cry it from the rooftops," he said. "This demolition of the old France … All manner of profanation, degradation, and ruin are all at once threatening what little remains of these admirable monuments of the Middle Ages that bear the imprint of past national glory. Though impoverished by revolutionary ravagers, by mercantile speculators, and above all by classical restorers, France is still rich in French monuments. The hammer that is mutilating the face of the country must be stopped.

The need for conservation had already been called for by Abbé Henri Grégoire, who in 1793–1794 argued for the liberating force of monuments and their documentary importance, principles which were shared by the National Convention. The concern with looting and demolition increased, partly due to the support of intellectuals such as Quatremère de Quincy, and the post of Inspector of Historic Monuments was created in France in 1830.

A similar cultural milieu also existed in Italy, as is shown by the Chirograph issued by Leo XII in 1825 to inspire restoration work on the basilica of San Paolo fuori le mura at Rome:

> firstly, we hope to satisfy the wishes of scholars (*eruditi*) and of those who admirably advocate the preservation of ancient monuments in the state in which they arose through the work of their founders. No innovation should be made in the architectural forms or proportions, and none in the ornament of the resurrected building except to

> exclude some small things that, in times after its original foundation, could have been introduced by the whim of later ages.

Meanwhile, Giambattista Piranesi (1720–1778) had promoted the taste for "ruins" throughout Europe.

In England, the taste for garden architecture increased the interest in ruins, which fuelled the taste for picturesque features in the landscape. The theme of ruins was exalted by Edmund Burke (1729–1797), who associated aesthetic values with the principles of conservation. However, the leading figure in conservation was John Ruskin (1819–1900), whose ideas attracted many admirers in Europe. A writer and intellectual with a passionate interest in painting and architecture, he championed the idea of the romantic, decadent ruin. His most important works on the theme of conservation and restoration were *The Seven Lamps of Architecture*, published in 1848, and *The Stones of Venice* (1851). He adopted an extremely radical stance in which there was a clear distinction between conservation and restoration; while the former was supposed to be restricted to structural interventions required to reduce dilapidation, the latter was a forgery of the monument: "It means the most total destruction which a building can suffer; a destruction out of which no remnants can be gathered: a destruction accompanied with false description of the thing destroyed".

Inspired by the principles developed by Ruskin, William Morris founded the "Society for Protection of Ancient Buildings" in 1877, which was involved in the conservation of many monuments, including St Mark's Basilica in Venice.

The French theory of stylistic restoration contrasted starkly with the English approach. Each monument in its (presumed) original state was deemed to constitute a "stylistic whole" which underpins the aesthetic quality of the work. The task of restoration was to restore this aesthetic quality by eliminating all alterations and completing it in its original style. This debate was an issue that was tackled by the first inspectors of monuments in France, in particular Ludovic Vitet and Prospére Merimée, although it was Viollet le Duc (1814–1879) who became its staunchest supporter and theoretician and, at the same time, its most prolific implementer. He left a profound mark on restoration in France and also in Italy. His theory is famously summarised in the following statement: "To restore a building is not to repair or rebuild it but to re-establish it in a state of entirety which might never have existed at any given moment".

This stance has influenced the cultural awareness of many architects up to the present day; they have treated ancient monuments as features of an artificial environment on which to intervene in order to leave their own indelible mark. Viollet le Duc was also a prolific theoretician and published the *Rational Dictionary of French Architecture* in ten volumes between 1854 and 1868. The most important project, which brought him everlasting fame, was the restoration of Notre-Dame, which was in a state of complete abandonment after the revolution and had been subjected to vandalism. He restored both the architecture and all the sculptures, with much rebuilding in the original style; he also designed and made the cast-iron spire (45 m. high) between the main nave and the transept, while he did not complete the two towers on the façade, as is shown in Figure 1.6. He also carried out interventions in the chapel of Sainte Chappelle, for which he constructed a wooden spire (33 m. high). Lastly, he was involved in the restoration of the ancient city of Carcassonne and, in particular, its castle; the restoration work, which was philologically arbitrary but extremely evocative and much criticised, transformed the city into a medieval setting for shows and film shoots (Figure 1.7).

Figure 1.6 Paris, Notre-Dame – Wikipedia

Figure 1.7 The historic city of Carcassonne – Wikipedia

There were many followers of Viollet le Duc in Italy, such as Luca Beltrami, who restored the sixteenth-century Palazzo Marino in Milan; and Alfonso Rubbiani, who restored the fourteenth-century Palazzo dei Notai in Bologna. Viollet le Duc's theories have always intrigued architects due to the intrinsic freedom they gave to their restoration projects, and they spread throughout Europe. However, they also led to "modern alterations, vaguely in keeping with the original style", which ill befitted the modern theory of conservation based on an innovative historical and scientific approach.

A particularly conservative standpoint was adopted by the Austrian architect Alois Riegl (1858–1905), who introduced the concept of *Kunstwollen* or "artistic intention". According to this notion, the monument took on relative artistic significance, which would now be ascribed to the dominant historic culture, while the historic significance was exalted. This vision led to the need to preserve monuments as a document since all monuments possess documentary value. Consequently, all arbitrary human intervention on the substance of the monument was to be excluded. The monument should therefore not undergo either additions or subtractions, nor the completion of what had become dilapidated due to weathering, nor the subtraction of what had been added, deforming the original form. It was an extremely modern approach from a historical perspective but theoretically radical from a conservational perspective.

The Italian approach to conservation was more complex and intricate. In 1807, Raffaele Stern (1774–1820) restored the eastern part of the Colosseum. He sensed "the fascination exerted by damaged and precariously balanced arches and secured them with bricks in a position as though they had been stabilised at the moment of collapse" (Marconi 1988), as shown in Figure 1.8. The intervention, which has been symbolically at the heart of ideas about restoration, was examined in both its conservative and innovative aspects. Indeed, "the intervention took on, in a paradigmatic way, the hypothesis of full respect for the monument extended, with mature critical detachment, both to the defence of the material substance and to the signs of the passing of time" (Carbonara 1996); other architects emphasised the use of brick and the imposing buttress as a sign of new times in which the building intervention is viewed almost as a comparison with antiquity.

Figure 1.8 Rome, the buttress built by Stern to stabilise the Colosseum – Wikipedia

Stern was commissioned by Pope Pius VII to restore the Arch of Titus on the Palatine Hill. The arch, which provides ancient evidence of the expulsion of the Jews from Palestine ordered by Titus in AD 79, has always been a fairly blatant symbol of anti-Semitism, and it required refurbishment and restoration. When Stern died, the work was unfinished but much of the material had already been prepared. The task of completing the work was undertaken by Giuseppe Valadier (1762–1839) who restored many monuments such as the Ponte Milvio at Rome and the cathedral of Urbino. He finished Stern's work on the Arch of Titus by installing columns, cornices and capitals, with substantial additions which displayed significant differences from the ancient ruins. Since then, debates have raged and the restoration has remained such a shining example that it was mentioned in the Italian Charter of Restoration in 1972. Regardless of the restoration of the Arch of Titus, in which Stern's imprint and materials emerge very clearly (as can be seen in Figure 1.9), Valadier, who carried out a much more cautious intervention on the Colosseum with the reconstruction of several arches, made an intriguing suggestion in a lecture given at the Accademia di S. Luca. He proposed that intervention on monuments should only involve "returning them to a healthy state" since the sole purpose of restoration was to ensure stability and improve understanding of the ancient work of architecture.

A radically different approach was pursued by Giacomo Boni (1859–1925), a leading admirer of Ruskin, with whom he had close ties. Boni developed his theory in a document "on the conservation of ruins and excavation finds" presented at a major conference – *Primo Convegno Italiano degli Ispettori Onorari dei monumenti e scavi* – held in Rome in 1912, in which he argued that "the authenticity does not represent the main quality of monuments,

Figure 1.9 Rome, the Arch of Titus – Wikipedia

but it is a necessary condition of any quality that they might have". The rules that were adopted and suggested were conservative forms of intervention that followed the rule of differentiation of the new from the old without creating pointless "false notes".

Giuseppe Fiorelli (1823–1896) was a particularly interesting figure within the Italian context. Although he worked in the archaeological field, some of his innovative ideas and guidelines have influenced the theory of conservation. Appointed Director of Excavations at Pompeii in 1860, he introduced a method of systematic excavation aimed at preserving the fabric of ancient buildings, eliminating the antiquarian-style approach that was akin to a treasure hunt. He worked on horizontal layers, proceeding from the top down, inventing "stratigraphic excavation", which only gained scientific recognition for all monuments in the second half of the twentieth century. He introduced the use of the "Excavation Diary" ("Giornale degli Scavi") to provide a detailed description of the location of the excavation, as well as the number and quantity of buildings and artefacts brought to light. He systematically published the results of the excavations and introduced the technique of making "plaster casts". In order to save frescoes and mosaics, he had them "detached" and kept in the Museo Archeologico Nazionale in Naples. This method, which was subsequently criticised for removing the artefact from its original context, did enable the gradual creation over time of a gallery of Pompeian paintings in the museum. When in 1875 he became Director General of Museums and Excavations, Fiorelli promoted his methods throughout Italy, contributing ahead of his time to the foundation of scientific archaeology. Using a thoroughly modern approach, he argued that "the conservation of monuments and artefacts will be deemed complete when it is preceded by the full-scale survey of everything that forms the subject of science without which it is impossible to decide what can be preserved and how to do so" (D'Angelo and Moretti 2004), implicitly criticising stylistic restoration and foreseeing the emergence of scientific research into material history.

The end of the century is marked by the figure of Camillo Boito (1836–1914), an architect, theoretician of restoration and writer. He was a champion of the Gothic revival, who initially espoused the ideas of Viollet le Duc, as can be seen in his intervention on Porta Ticinese in Milan. His writings on restoration reveal his support for conservation and an utter respect for the monument as a piece of historical evidence, as well as the need to restrict intervention to ensuring stability and, where necessary, to repairs with materials that indicate their modernity. He had immense cultural influence, although, in his judgements in architectural competitions, he displayed a degree of ambiguity by supporting the demolition of historic districts. He also made some innovative proposals at the Congress in 1883 that led to the first Italian Charter of Restoration, which suggested that monuments should be consolidated rather than repaired and, if necessary, should be repaired rather than restored, avoiding additions and innovations. Boito argued for the differentiation of materials, the inscription of the date of the restoration on each added piece, a report and photographic recording of the work carried out and a copy of the documents deposited with the authorities. Unfortunately, this last recommendation has often been largely ignored.

The ideas put forward by Gustavo Giovannoni (1873–1947) lead straight into the debate that raged in the twentieth century. He highlighted the difficulties of restoration:

> and we know the balance that the restorer is required to display and the determination to resist the compelling desire to complete what is incomplete ... During restoration, it is easier to act rather than refrain from acting: it is extremely difficult to impose the limits and forms of what may not be a reasonable reconstruction.

He added:

> we have sought to carry out honest and sincere work which is revolutionary in historic and artistic terms, far removed from the cold concepts of die-hard conservators who will not even accept the repair of a small part of a destroyed cornice or corroded face, as if they were the dangerous attempts of aesthetes who, through deductions and analogies, would like to complete each part of a monument.
>
> (D'Angelo and Moretti op. cit.)

Giovannoni defined five types of restoration: consolidation, reconstruction, liberation, completion and innovation. The conservative restoration undertaken between the 1920s and 1940s was inspired by the work of Stern and Valadier in the use of traditional materials and buttresses, which sometimes included the use of concrete. Thus, Canina reconstructed the arches of the inner circle of the Colosseum in brick; Calza Bini and Fidenzoni constructed the buttresses of the Theatre of Marcellus, also in Rome (Figure 1.10); Maiuri added a load-bearing wall using blocks of equal length (*opus isodomum*) to the fortifications of the Acropolis of Cumae. Jole Marconi Bovio used bricks and cement for the rear wall of the cavea of the Theatre of Segesta, blocks of local stone and original blocks for the cavea and concrete in the anastylosis of Temple E at Selinunte. The relative continuity of tradition was broken by the Charter of Restoration of Athens of 1931; together with reinforced concrete, it was considered opportune to insert "injections of concrete, tie beams, trusses and steel frameworks, since these preconceptions had been overcome and the constructional elements were viewed as useful features". This was undoubtedly due to the lack of knowledge of the properties of the new materials, to the diminished importance of monuments as a historical document and, lastly, to a lack of sensitivity to the material history of antiquity. The use of concrete began to spread, like a panacea, to numerous monuments. Article V of this charter also recommended "concealing interventions made of reinforced concrete";

Figure 1.10 Rome, Theatre of Marcellus – Wikipedia

this expedient can be seen in emblematic fashion in the reconstruction of the Capitolium in Brescia carried out between 1938 and 1943 by Laurenzi, who constructed columns and structures made of reinforced concrete covered in brick, shown in Figure 1.11. Meanwhile, the use of reinforced concrete had spread throughout the Mediterranean with the interventions of Balanos at the Parthenon and Luigi Pernier in Crete, while Laurenzi used reinforced concrete to strengthen the columns of the temple of Pythian Apollo on Rhodes, a detrimental practice that spread rapidly and caused serious damage. The use of reinforced concrete rapidly spread to the restoration of all types of monument from churches to palaces, becoming a technique of intervention that caused irreversible damage to the world's historical heritage. The use of reconstruction, better known by the term *anastylosis*, generally concerns archaeological restoration. It normally consists of reconstruction using excavation material. Reconstruction is usually partial, leading to the problem of "lacunae" and the definition of the missing pieces in each specific case (D'Agostino, Martines 1997). Brick, a material with almost limitless durability which goes well with other stone materials, has been usually used to fill gaps and, as has been seen, to cover additions made using reinforced concrete.

Restoration through the isolation of the monument and the elimination of subsequent additions was used extensively in the first half of the twentieth century. It led to major demolition work, for example in Naples to isolate Porta Capuana (Figure 1.12) and in Milan to isolate the colonnade of San Lorenzo (Figure 1.13) as well as the complete demolition of districts – frequently medieval ones – in many Italian towns and cities.

Restoration by completion tends to give a finished form to a monument. Examples include several cathedrals such as the cathedral of S. Maria del Fiore in Florence and the Duomo in Naples, (Figure 1.14). From the perspective of conservation, an extremely opportune

Figure 1.11 Brescia, Capitolium – Wikipedia

Figure 1.12 Naples, Porta Capuana – Wikipedia

Figure 1.13 Milan, the Columns of San Lorenzo – Wikipedia

Figure 1.14 Naples, the Duomo – Wikipedia

case of restoration by completion was the reconstruction of the roofs at Pompeii where, for example, Maiuri reconstructed the roofing of the atrium of the House of the Vettii and its peristyle (Figure 1.15).

Restoration by renovation seeks to add parts to the monument that have been "identically" formed for practical reasons, to enlarge the monument and to "complete it". It does not differ greatly from restoration by completion, and these types of restoration are obviously the cause of fierce debate because they detract from the authenticity of the monument. The construction of the twin towers on each side of the Pantheon, constructed in the seventeenth century and demolished in the nineteenth century, is a classic example.

Piero Sanpaolesi (1904–1980) worked in the first half of the century. He was extremely active from the 1930s onwards, heavily involved in the administration of cultural heritage as Superintendent, and later became professor at Pisa and later at Florence where, in 1960, he set up, within the university, the Institute for the Restoration of Monuments (*Istituto di Restauro dei monumenti*), leaving a large number of pupils who espoused his approach. He promoted the constructive study of monuments and the various forms of deterioration, supporting radical conservation. In his anxiety to experiment with new materials for consolidation, he adopted fluorosilicates, which did not achieve the desired results, although, unfortunately, they continued to be used for a long time. His numerous restoration projects included the Triumphal Arch of Laurana at Castel Nuovo in Naples (Figure 1.16), the Ca d'Oro in Venice (Figure 1.17) and the façade of Palazzo Rucellai in Florence.

Figure 1.15 Pompeii, peristyle at the House of the Vettii – Wikipedia

Figure 1.16 Naples, Triumphal Arch of Castel Nuovo built by Laurana – Wikipedia

Figure 1.17 Venice, Ca' d'Oro – Wikipedia

This brings us to the heart of the twentieth century, ravaged by wars, earthquakes and uncontrolled consolidation (Lizzi 1981), but also enriched by a new cultural perspective which tended to highlight the importance of the monument as document of historic built heritage, championing conservation as an archive of the material history of humanity. During the twentieth century, as the concept of the conservation of cultural heritage in its integral form became increasingly well established, restoration became an operation that sought to restore the integrity of the historical and artistic monument as far as possible. In the first half of the twentieth century, there were two conflicting theories at an international level. The first theory, which emerged in the United States (Pease 1950), argued for the primacy of scientific research in the distinction between original material and subsequent additions, while the second theory, which emerged in Italy and was formulated by Cesare Brandi, provided a critical reflection on cultural heritage of an aesthetic and historical nature, while acknowledging the intrinsic value of technical and scientific data. Brandi's vision of restoration as a "critical act" immediately attracted widespread international support (Brandi 1997). His work culminated in the foundation of the Central Institute for Restoration (*Istituto Centrale per il Restauro*), which was opened in Italy in 1939 and conducted restoration projects that also gained considerable international renown. Some of the most interesting initiatives included the scrupulous application of methods for painting, used by Paolo Mora in the restoration of the central architrave of the Duomo of Siena, carved by Tino da Camaino. It was interpreted as a polemical response to the large-scale intervention ordered by André Malraux for the historic façades of Paris, which had been turned white by rash cleaning operations using industrial sandblasting and water. Meanwhile, the importance of technique, already formulated at a philosophical level by Martin Heidegger and taken up in Italy by Giovanni Urbani (Urbani 2000), emerged in all its complexity and was reflected in the humanities in the creation of material history, and in the scientific field by the creation of archaeometry.

1.4. Material history and scientific research

During the second half of the twentieth century, the need for conservation grew considerably, due to the emergence both of material history and archaeometry.

In 1949, March Bloch (Bloch 1949) published his famous essay that put forward the concept of history as the science of humanity and gave historical research a decidedly scientific character. Simultaneously, the creation of the magazine Les Annales and the work of historians such as Fernand Braudel and Le Goff extended, in the wake of Bloch's pioneering work, the field of historical research, which now included all human activities and, in particular, led to the creation of material history. This became the global approach to historiography from the history of economics to the history of clothing. In Italy, for example, it is worth mentioning the essay by Giuseppe Galasso, "*Nient'altro che storia*" (Galasso 2000), and his groundbreaking essay "*La storia e le storie*" (Galasso 2018). From this cultural perspective, it is clear that the very material from which an object is made can become a historical document, and this is particularly true of cultural heritage. This is how the concept of the monument as document, which evolved in the French cultural milieu, should be interpreted. This overturning of the concept is quite clear; each monument, with its own materials, therefore constitutes a document from the archive of the material history of humanity. Simultaneously, the sub-discipline of archaeometry emerged, propelled by British and American culture, and as part of the unbridled scientific and technological development of the twenty-first century. Archaeometry, also known as archaeological science, consists of the scientific and diagnostic study of the materials of cultural heritage; it concerns physics, chemistry, biology, geology and computer science. Unfortunately, it excluded engineering which, despite playing a key role in the safety and conservation of works of art, struggled to gain awareness of its own role, which would only emerge in all its complexity at the beginning of the new century (D'Agostino 2017). Archaeometry would soon make its crucial scientific contribution to all the many different materials that constitute cultural heritage, such as paints, bronzes, glass, fabrics, stone and ceramics. In 1958, Oxford University began publishing the *Bulletin of the Research Laboratory for Archaeology and the History of Art*. As of 2001, the journal has been published on behalf of the Research Laboratory for Archaeology and History of Art, Oxford University, in association with the Gesellschaft für Naturwissenschaftliche Archäeologie Archäeometry and the Society for Archaeological Sciences by Wiley-Blockwell. The impact of archaeometry would become increasingly important in historical and scientific research. Its contribution to the dating of finds is just one obvious example. Already in an advanced state of application, thermoilluminescence dating would have a major impact on the dating of masonry, leading to major advances in historical and cultural studies. The fields of architecture and engineering should be displaying particular interest in this field, although they are currently on the margins of scientific research that could shed light on the history of ancient monuments, providing a major contribution to their conservation.

References

Bloch M., [1949], *Apologie pour l'historie ou mètier d'historien*, A. Colin, Paris.
Brandi C., [1997], *Teoria del Restauro*, Piccola Biblioteca Einaudi, Turin.
Carbonara G., [1996], Teorie e metodi del restauro, in Carbonara G., (ed.), *trattato di restauro architettonico*, UTET, Turin.
Cordaro M., [2000], *Restauro e tutela*. Scritti scelti (1969–1999), Annali dell'Associazione Bianchi Bandinelli no. 8.
D'Agostino S., (a cura di), [2017], *Ingegneria e Beni Culturali*, Il Mulino, Bologna.
D'Agostino S., Martines R., [1997], *Reintegrazioni architettoniche e reintegrazioni strutturali: dialogo sul tema della conservazione*, Atti XIII Conv. Scienza e Beni Culturali, Bressanone 1997.
D'Angelo D., Moretti S. (a cura di), [2004], *Storia del restauro archeologico*, Alinea, Florence.
De Maria S., [1988], *Gli Archi Onorari di Roma e dell'Italia Romana*, L'Erma di Bretschneider, Rome.
Galasso G., [2000], *Nient'altro che storia*, Il Mulino, Bologna.
Galasso G., [2018], *La storia e le storie*. Atti VII Conv. Int. Conf. Di Storia dell'Ingegneria, Cuzzolin, Naples.
Lizzi F., [1981], *Restauro Statico dei Monumenti*, SAGEP, Genoa.
Marconi P., [1988], *Dal piccolo al grande restauro*, Marsilio, Venice.
Melucco Vaccaro A., [1989], *Archeologia e Restauro Il Saggiatore*, Mondadori, Milan.
Mertens D., [1984], *Aspetti dell'Architettura a Crotone*, in Crotone, Atti XXIII Conv. di Studi sulla Magna Grecia, Taranto.
Pease M., [1950], *The Future of Museum Conservation*, Wiley Online Library.
Pergoli Campanelli A., [2015], *La nascita del restauro*, Jaca Book, Milan.
Schmidt K., [2000], *Göbekli Tepe and the Rock of the Near East*, TüBa-Ar, 3.
Urbani C., [2000], *Intorno al restauro*, Skira, Milan.

Chapter 2

Conservation and restoration in Europe

During the second half of the twentieth century the theme of conservation attracted considerable international interest, leading to the establishment of various institutions. As early as 1945, UNESCO, the organisation for education, science and culture, was set up by the United Nations. Its activities include protecting sites of special historic and artistic interest throughout the world. In 1956, the general conference of UNESCO promoted the foundation of the International Centre for the study of the preservation and restoration of cultural property (ICCROM) which was established in Rome in 1959 and has undertaken numerous scientific missions and set up many training courses. Lastly, following the adoption of the Venice Charter for restoration in 1964, the International Council of Monument and Sites (ICOMOS) was set up in 1965. All these institutions share an identical vision of conservation.

In 1964, the British Standards Institution defined maintenance as actions which "retain an item in, or to restore it to, a state in which it can perform its required function", and in 1990 it was emphasised that regulations should be based on a comparison between needs and available resources. The Nara Conference held in 1994 formulated a resolution on the issue of authenticity which makes it necessary to respect the cultural and social values of each country:

> All judgements about values attributed to cultural properties as well as the credibility of related information sources may differ from culture to culture, and even within the same culture. It is thus not possible to base judgements of values and authenticity within fixed criteria. On the contrary, the respect due to all cultures requires that heritage properties must be considered and judged within the cultural contexts to which they belong.

At the conference, held in S. Antonio (Texas), the concept of authenticity was extended not just to material integrity but also to the cultural identity of peoples.

Lastly, at the conference on cultural heritage held in Ljubljana in 2008, considerable attention was focused on maintenance viewed as an activity designed to minimise dilapidation by avoiding, or at least delaying, restoration in order to preserve the material authenticity and integrity of the artefact. Emphasis was also placed on users and their training as the promoters of the culture of maintenance. Obviously, these general concepts take on specific connotations for different countries, and the cultural perspective and institutional framework of leading European countries are briefly summarised below.

DOI: 10.1201/9781003160960-2

2.1. Conservation and maintenance in the UK

The British model reflects the intrinsic empiricism of British culture. There is less state intervention and relatively limited public ownership, while private enterprise plays a larger role, operating in both individual and collective forms. The following sectors are related to conservation: cultural heritage, which includes archaeological and historical sites, property of art-historical interest and the landscape; the cultural institutions which, operating within a humanistic and scientific cultural framework, are concerned with museums and libraries; and, lastly, the arts and media involved with theatre and music and, more generally, the media. Legislation differs according to the sector. For example, cultural heritage is mainly under the remit of local authorities, while the government is responsible for planning and defining the guidelines. This has led to the collective management of protected heritage, which is listed in special registers.

The administrative structure is divided into a series of institutions. The Department of Culture, Media and Sport is concerned with the planning and definition of rules and best practices. The Department of the Environment, Heritage and Local Government formulates policy for safeguarding heritage. English Heritage, which was set up in 1983 by the National Heritage Act, is the body that connects central and regional power. It has numerous functions such as funding, safeguarding heritage and developing guidelines. It is supported by the Historic Buildings and Monuments Commission and is organised in regional offices. Since 2002, it has been authorised to sell guidebooks and souvenirs, which are protected by copyright. The National Trust, which was founded privately in 1895, was incorporated in 1907 by the National Trust Act, which confirmed its status as an association whose members provide financial resources. It has a wide-ranging remit in safeguarding architectural, landscape and art-historic heritage.

Public intervention is restricted to allocating funds through the Council of England. Funding is obtained from the National Lottery or from companies that are entitled to tax relief. The foundation of Maintain Our Heritage, a non-profit organisation, in 1999 was an extremely important initiative. It encourages maintenance intervention, in particular with regard to extraordinary maintenance. The association plays an extremely important role because it follows the intervention from feasibility studies to implementation. Once the intervention is complete, forms are prepared for the various parts of the building, indicating the maintenance times and arranging inspections. Unfortunately, the association has no government funding.

2.2. Conservation and maintenance in Germany

In the period following the Second World War, Germany sought to downsize the role of the Weimar Constitution of 1919 by creating the Basic Law for the Federal Republic of Germany (*Grundgesetz*). It delegates responsibility for cultural heritage to individual *Länder* (states), each of which has its own constitution. Only in the late 1960s, following the Venice Charter of 1964, did the *Länder* define legislation and administrative apparatuses for safeguarding heritage. The building law has federal significance and sets out criteria for safeguarding areas and monuments of historical interest. However, there is a law for the protection and conservation of monuments, which involves authorisation and limitations for intervention on buildings.

Despite the differences between the legislative systems, they all share a definition of cultural heritage that includes

> the movable and immovable material heritage that has artistic, historical or archaeological significance, or that constitutes the distinctive features of a territory, or a town or city whose conservation is necessary for artistic, historical or scientific reasons or for promoting historical and national awareness which encourages public interest in this field.
>
> (Totaro 2010)

Safeguarding heritage is the responsibility of each state in which the heritage is situated, with the possibility of delegating to municipalities, while their protection comes under the autonomous jurisdiction of local authorities. Since 1999, a single ministry has existed which brings together federal jurisdiction; it runs the library and museum system and sets up foundations which involve the joint participation of Federations and *Länder*. The relationship with the private sector is regulated by the Council for Cultural Heritage (*Denkmalrat*), which drafts contracts between the public and private sectors. In economic terms, public funding is available for restoration and conservation.

During the late twentieth century, Germany became increasingly aware of preventive conservation, and numerous associations were set up.

The Monument Service (*Monumentendienst*), which was set up in 2004, is sponsored by the Federal State for Lower Saxony; the association Altbau und Denkmalservice Gesellschaft zur Erhaltung des Kultures besand V. was founded in 1999 by the Federal State of Hessen. These associations are inspired by the Dutch model, which provides a recording service that illustrates the state of the building and suggests the required intervention. In 2004, all the associations became part of the Federal working group of the inspection services for monuments and historic buildings, financed with regional funds, with the intention of standardising recording practices and intervention criteria.

2.3. The Spanish model

The Spanish Constitution of 1978 states that

> public powers guarantee the conservation and promote the recognition of the historical, cultural and artistic heritage of the peoples of Spain and the heritage that brings them together, whatever the legal system and its ownership. The criminal law will impose penalties on attacks carried out against this heritage.

Interestingly, Spain is the only European country to invoke criminal law.

The Spanish Historical Heritage Act, passed in 1985, makes a distinction between three different levels of heritage and protection: the Bienes de Interes Cultural (BiC), regulated by the Civic Code; the movable heritage included in the General Inventory (Inventario General) due to its significance; and, lastly, a third category whose relative importance is recognised. The 1990s witnessed the decentralisation of the protection of cultural heritage from the state to Autonomous Communities. This process has led to greater autonomy for Autonomous Communities which have their own constitutions. The Autonomous Communities play a crucial role in the conservation of historic heritage, even though they are subordinate to state legislation. The Institut Català de les Industries Culturals (ICIC) is at the forefront of cultural heritage policy; in particular, its remit covers "cultural industries", in other words firms that produce, distribute and sell "cultural products", encouraging cooperation, agreements and promotion.

In economic terms, the state has developed numerous instruments to encourage the conservation and management of cultural heritage: firstly, the allocation of 1 per cent of public works funded by the state for the conservation and enhancement of historic heritage; tax relief for private individuals and companies; exemption from heritage tax for cultural property that forms part of the historic heritage; tax incentives designed to encourage patronage; and tax relief and allowances for the costs of conservation, restoration and renovation, including the restoration of façades and roofs.

2.4. The French model

It was during the French Revolution, in the midst of the Enlightenment, that France began to draw up measures for safeguarding historic heritage. This process gradually led to the law passed in 1913, which acknowledged public interest in public heritage. The Code du Patrimoine was published on 24 February 2004 and has been systematically updated. It is arranged in seven parts that deal with general regulations, archives, libraries, museums, archaeology, historical monuments, the landscape and regulations related to overseas territories.

In terms of conservation, cultural heritage is subject to the process of classification which is structured on successive levels and concludes with registration in the mortgage register.

> The state exerts permanent control over classified heritage which concerns the terms of use and conservation. While public ownership is subject to the positive obligation of ensuring the conservation of the object, private owners are only required to observe negative behaviour; in other words, they should avoid any form of alteration, repair or restoration without the necessary authorisation.
>
> (Totaro, op. cit)

Once authorisation is obtained, public or private owners must carry out the work under their control and must undertake what is deemed necessary by the state. Buildings situated within a perimeter of 500 metres around a safeguarded historical building cannot undergo alterations without administrative authorisation.

There is a supplementary inventory which contains cultural property of lesser importance. In the case of maintenance and restoration, state funding of up to 40 per cent of the total cost is available. For the investigation of *Classements*, the Ministry of Culture and Communication makes use of the Commission National des Monuments Historiques. The Ministry is divided into Directorates-General: the Direction Général des Patrimoines is responsible for the maintenance, conservation, restoration and enhancement of monuments. The Ministry is divided into Directions Régionales des Affaires Culturelles (DRAC), which are subordinate to the prefect and structured in various responsibilities delegated to the cultural policy in the Region, ranging from conservation and enhancement to development policies with a very broad sphere of action. Another heritage institution is the

Centre des Monuments Nationaux, which carries out the maintenance, conservation and restoration of national monuments, while the Fondation du Patrimoine has the task of both safeguarding heritage and training in the fields of restoration and enhancement. There are also foundations with specific aims. State intervention is also available in the form of funding and grants, but both the Departments and the Communes can allocate funds from their budgets, financed partly with the proceeds of the sale of tickets to historical monuments.

2.5. The Italian model

Italy has been at the forefront of safeguarding historic heritage, beginning with legislation passed by the different states prior to unification, starting with the pre-unification states. In the period following Italian unification, the protection of heritage was entrusted to the Ministry of Public Education, which had a special department for monuments and fine arts. After the Second World War, article 9 of the Italian Constitution drafted in 1947 stated that "The Republic promotes the development of culture and scientific and technical research. It safeguards the landscape and the historical and artistic heritage of the nation".

Due to the postwar reconstruction programme, which had had devastating effects on the nation's cultural heritage, the Italian cultural elite began to campaign for the establishment of a Ministry for Cultural Heritage. This led in 1974 to the foundation of the *Ministero per i Beni Culturali e Ambientali* and, as from 1998, the *Ministero per i Beni e le Attività Culturali*. Its sphere of action was extended to the theatre and cinema and even to tourism, which has recently been separated. The Ministry operates on the territory through the *Soprintendenze* (Superintendencies), initially divided into archaeological, architectural and art-historical *Soprintendenze* with relevant effects on the monitoring of the territory. The *Soprintendenze* express a binding opinion on all conservation and restoration projects, supervising them in the implementation stage. Unfortunately, the ministerial budget is extremely limited, probably one of the lowest in Europe, Moreover, the Ministry has an endemic shortage of staff. Besides the Ministry, there is also the *Fondo per il Culto*, the legacy of completely outdated legislation, which is part of the Ministry of the Interior and supervises almost a thousand ecclesiastical buildings. The situation is made even more complicated by the widespread presence of ecclesiastical property.

Given the widespread risk of earthquakes throughout Italy, engineering regulations have become imperative. After formulating a praiseworthy Code for Cultural Heritage, the *Ministero dei Beni Culturali* issued guidelines for intervention in seismic zones and drew up a risk charter. However, there has recently been a complex problem between conservation and safety which has led to a dispute between engineering – the science of action and the possible – and the judiciary, which lays claim to an arcane principle of absolute safety.

2.6. The Greek model

The case of Greece is particularly interesting since the nation is the custodian of a culture and monuments which constitute the cornerstone of Western civilisation. As is well known, the most famous Greek monuments have undergone serious ordeals over the centuries, in particular the monuments of the Acropolis, with religious settlements, the shelling by the Venetian republic and the plundering by Lord Elgin. After it gained independence, Greece was finally able to turn its attention to the past. When Otto I (1833–1862) came to the throne on 25 January 1833, a cultural movement emerged that was marked by classicism, exalted by the ancient past. Karl Friedrich Schinkel, the great German architect and promoter of classical architecture almost in the image of antiquity, even designed the construction of a royal palace on the Acropolis, a project which, fortunately, is famous but was never put into practice. Work was carried out during this period, and the bastions of the Propylaea were demolished.

Between the late nineteenth century and the early twentieth century, there was a desire to reconstruct the initial elegance and clarity of the forms and profiles of the monuments – in other words, their essential classical appearance. The radical application of this criterion led to the demolition and removal of all the buildings built on the Acropolis during the period of Turkish domination. This period was dominated by Nikolaos Balanos (1869–1943), a Greek architect who, after studying at the École Nationale des Ponts et Chaussées, became Director of the Technical Department of the Greek Ministry of Public Instruction. He initially worked on the consolidation of the monuments using *anastylosis*, a term that refers to the reconstruction of ruined buildings with original materials and techniques, although subsequently, as in other countries, he employed contemporary building materials (steel and concrete), which over time proved harmful to conservation. During the 1950s, the idea that cultural heritage could be exploited economically, supported by renewed tourism, led to several instances of negligent intervention and, even though the war did not prove to be particularly harmful, significant damage was incurred by the earthquakes of 1953, 1956 and 1965. The return to democracy in 1974 proved to be the starting point of political and social maturity, encouraged by the entrance of Greece into the European Economic Community. The changes affected the safeguarding of heritage, the type of heritage and structure that required safeguarding, information and education, the legislative framework, authorities, funding and the procedures for carrying out the work. Safeguarding heritage was enshrined in the constitution of 1975 and the amendments made in 1986 and 2011.

Legislation designed to safeguard antiquity and cultural heritage was promulgated in 2002, automatically protecting all cultural heritage (material and immaterial, movable and immovable), historic and archaeological sites up to 1830, the year of the foundation of the Greek state, while later works were safeguarded after applying general legal restrictions. A structure was set up that had parallels with Italy; the Ephorias (directorates) played a similar role to the *Soprintendenze*. The law also established Local Councils for Monuments for each administrative region with markedly interdisciplinary criteria for safeguarding heritage, involving the participation of the scientific community. Since 1975, the Ministry of Culture has been backed up by the Ministry of Territory, Environment and Public Works for safeguarding historic centres and, since 1984, the landscape and more recent monuments. Other bodies include the Central Archaeological Council and the Central Council of Monuments with a wide range of staff comprising officials, superintendents and university lecturers and professors. The councils are required to give their opinions on many issues such as compulsory purchases, demolition and exports.

The overlapping responsibilities of many supervisory bodies represent a weak point of the Greek approach, while the work of the Commission for the Conservation of the monuments of the Acropolis has been exemplary, acting as a "school" for archaeological restoration. Towards the latter part of the twentieth century, there was a trend towards the systematic recording of buildings, which attributes great importance to the original material and its study. This approach takes account of the complexity and historical stratification of monuments and supports recognition of the distinctive phases, and it emphasises the educational significance of the monument, suggesting the display of the restoration documentation.

Lastly, it supports the reversibility of the intervention, providing future generations with the opportunity and right to return to the same operations.

The legislation has had the advantage of addressing all historic built heritage and extending "protected use" to housing, an effective measure for urban reorganisation which has, for example, led to the regeneration of the Athenian neighbourhood of Plaka.

The Ephorias, especially those responsible for safeguarding modern heritage, are experiencing some serious problems: a legal framework with many shortcomings, deficiencies in terms of structure and organisation, staff that are insufficiently skilled to cope with the enormous requirements, a lack of funding and resources and incentives which are either ineffective or, for private individuals, almost non-existent.

Unfortunately, the approach to conservation is marked by the utter lack of a concept of "maintenance" which requires highly skilled and committed staff. Maintenance is still not provided or organised while it ought to be an institutional requirement.

Since the state passes legislation but does not have an adequate structure nor the necessary vigour or economic resources, conservation remains an extremely problematic issue.

2.7. The Portuguese model

Conservation and restoration in Portugal have parallels with the situation in Italy. Like Italy, Portugal formalised its principles of conservation in the Constitution of 1989. A law passed in 2001 has included programmed maintenance as an objective for safeguarding heritage and has identified heritage as the totality of tangible and immaterial heritage that provides evidence of civilisation and has significant national interest. Portugal has also adopted a methodology similar to the Italian Risk Charter for planning an intervention strategy and defining maintenance plans. Conservation is the task of the state, while raising awareness and enhancement are functions divided between the state, autonomous regions and local authorities. Two ministries are responsible for heritage conservation – the Ministry of Culture and the Ministry of the Environment and Territorial Planning – and their structure is fairly similar to the equivalent Italian ministries (Totaro 2010, op. cit.).

2.8. The experience of the Netherlands and Flanders: Monument Watch

The crucial and innovative experience in terms of heritage in the Netherlands is Monument Watch (Monumentenwacht), which was founded in 1973 and is based on the interaction between technicians and users for the maintenance of built heritage. Since 1991, Monument Wach has played an active role in Flanders (the Flemish-speaking part of Belgium) as well. This independent organisation monitors buildings through regular inspections and maintenance. Its administrative headquarters are situated in Antwerp, with a permanent council that runs and coordinates the activities by ensuring the uniformity of operational practice. Inspections are carried out by two teams, one with responsibility for architecture and the other for interiors; the architectural team consists of an engineer or an architect and a worker with specialist skills in small-scale maintenance. Both teams draft a final report. This ensures the safeguarding of heritage, constant monitoring and cultural advancement of users who acquire the concepts of dilapidation and maintenance.

90 per cent of the activities of Monument Watch are state funded while money for the remaining 10 per cent comes from private subscriptions and the proceeds from inspections.

Additional funding comes from the National Lottery, which finances projects for heritage conservation. The inspections are supposed to take place every 2 years, although they are often carried out only every 3–4 years, due to lack of funding. Listed buildings are entitled to incentives such as limited reimbursement for expenses while, for private buildings, the state allocates up to 40 per cent of the expenditure approved by the government; the figure for churches varies from 60 to 90 per cent. There are no grants for unlisted buildings, although some cities provide funding for maintenance work, in particular for the restoration of façades. Nevertheless, maintenance is becoming more widespread, and many private owners are members of the association. Unfortunately, when owners undertake extraordinary maintenance, there is no longer any contact with the technicians of Monument Watch, which leads to a significant loss of information.

Monument Watch has achieved considerable popularity among owners and has led to an increase in regular maintenance with considerable advantages in terms of conservation.

2.9. Comparing European contexts

The concept of cultural heritage has become a crucial factor throughout Europe, although it takes on different guises in each country. While cultural heritage is a fundamental part of the structure of the country, the classification of property differs according to each nation. In every country, the state has duties in terms of the protection and management of cultural heritage, even though it is often included in federal legislation, as in Germany and Spain, where there is a lack of administrative uniformity. The methods and bodies for cataloguing heritage vary from nation to nation, as do the systems of funding.

With regard to maintenance, the method of inspections by consultants who carry out fact-finding research and small-scale intervention maintenance seems to play a crucial role. In Italy, a change of approach is required that ought to acknowledge the key role of ongoing and programmed maintenance, both for public heritage under state supervision and for privately owned heritage through the intervention of specific associations. Engineering could play a highly significant role in the development of this cultural mindset. Some monuments of outstanding importance are entrusted to institutions with special responsibility for the maintenance of religious buildings, known as *Fabricerie*. Important examples include the Veneranda Fabbrica del Duomo di Milano and the Fabbrica di San Pietro for the Vatican.

In terms of the financial burden, nearly all European legislation provides tax relief for private owners for the expenses related to maintenance and restoration. European policy tends to encourage the involvement of citizens and to support cultural associations, albeit with meagre resources.

2.10. Concluding remarks

This brief overview of the cultural heritage policies in several European countries leads to a series of conclusions about the importance of the conservation of cultural heritage in which the roots and cultural history of the continent can be identified. Approaches differ, not so much in terms of culture but in terms of administration and management. Given the current political climate, it seems unlikely that the European Union will be able to introduce uniform legislation for the protection and conservation of cultural heritage. However, it is vital that the underlying principles are largely identical and that Europe makes a contribution, providing funding and financing projects for conservation and restoration, especially in countries within the Mediterranean area. The situation varies due to the differences in seismic risk, which is particularly significant in Italy, Greece and Portugal.

Maintenance is a cornerstone of conservation in all countries, in keeping with the material culture of the historic heritage. Ideally, engineering will play a greater role in safeguarding cultural heritage.

In terms of the approach to maintenance, there are considerable differences. The extreme cases can be found in Belgium and the Netherlands, with the scrupulous programming of Monument Watch; and in Italy, where the concept of planned maintenance is struggling to be transformed into efficient operational programmes. With regard to economic management, there is a reasonable degree of uniformity of principle, with states taking responsibility for safeguarding public heritage and the provision of tax relief and incentives in favour of the conservation of privately owned heritage.

The importance of cultural heritage and its conservation is therefore recognised, although there is still a lack of economic and human resources which, in cases such as Italy, can become a major disadvantage.

References

Bilancia P., [2005], *La valorizzazione dei beni culturali tra pubblico e privato. Studio dei modelli di gestione integrata*, Franco Angeli, 2005.

British Standard no. 3811, [1998].

Buoso E., [2008], *I caratteri fondamentali della disciplina dei Beni Culturali in Germania in una prospettiva comparatistica.* In rivista giuridica di urbanistica no. 1 | 2, 2008.

Cecchi R., Gasparoli P., [2010], *Attività di prevenzione e cura su un patrimonio di eccellenza: il caso delle aree archeologiche di Roma e Ostia antica*, Conference Proceedings Scienza e Beni Culturali, Bressanone, 2010.

Della Torre S., [2010], *Preventiva, integrata, programmata: le logiche evolutive della conservazione*, Conference Proceedings Scienza e Beni Culturali, Bressanone, 2010.

Iadicicco Spignese S., Jiurina L., [1999], *La manutenzione come programmazione della conoscenza*, Conference Proceedings Scienza e Beni Culturali, Bressanone, 1999.

Interaunican Symposium on Authenticity in the Conservation and Management of Cultural Heritage, 1996, S. Antonio, USA, 1996.

Marconi P., [1993], *Il restauro e l'architetto*, Marulio, Venice.

Regione Lombardia, [2003], *Linee guida per la conservazione programmata del patrimonio storico architettonico*.

Rinaldi L., [2009], *Manutenzione vs. conservazione? Il cantiere del Duomo di Milano negli ultimi trenta anni*, Conference Proceedings Scienza e Beni Culturali, Bressanone, 2009.

Totaro G., [2009–2010], *Attività di manutenzione e cura sui Beni Culturali Architettonici: strategie e politiche di incentivazione*, Degree Dissertation, Politecnico di Milano.

Chapter 3

Construction in antiquity

The original phases of the development of human activity were undoubtedly inspired by the observation of nature and, like nature, they varied according to the different parts of the planet. Whether mountains with their caves or majestic trees that stand up to storms, careful observation of different territories provided humans with the criteria and modules that were essential for early human evolution (Figure 3.1). Our knowledge of the remote past is steadily increasing with surprising new discoveries such as the site of Göbekli Tepe, which developed between the twelfth and eleventh millennium BC, revealing the capacity of humans to transport huge weights and work stone with great sophistication and elegance (Figure 3.2).

In Mesopotamia, where the use of mudbrick predominated, social life emerged during the fourth millennium BC, which revolved around both large complexes (Figure 3.3) and farmers' houses (Figure 3.4).

Egypt developed its own autonomous culture, which was rich and complex, while Minoan and Mycenaean cultures were organised on a more human scale (Figure 3.5).

The development of architecture would be marked in the western hemisphere by a culture of land and rock, using fired brick and lime. Together with natural stone, these were the characteristic materials used for construction in Europe until the end of the eighteenth century.

However, a great stride forward in human intelligence took place with the origins of geometry (Figure 3.6), which made it possible to assimilate and reinterpret natural forms by creating walls, pillars, columns, arches and vaults; all these elements can be found in the spatial-volumetric approach to architecture.

3.1. Building practices

The ancient concept of construction therefore has an essentially experimental basis which stems from the observation of nature and is rationalised by geometry. Like many human activities, such as agriculture, husbandry, clothing and war, it is handed down through an experimental process that consists of *specific building practices*: the methods and procedures defined by complex experiences confirmed by practice that may go back hundreds of years. This has remained an innate, vibrant feature of rural life as well as the rich and complex world of handicrafts. When artisans achieve mastery of a craft and through their work provide evidence of the feelings of humanity, they become artists.

DOI: 10.1201/9781003160960-3

Figure 3.1 Vieste, the natural arch in Baia San Felice – Wikipedia

Figure 3.2 The site of Göbekli Tepe – Wikipedia

Figure 3.3 The Palace of Knossos – Wikipedia

Figure 3.4 Buchau in Württemberg, reconstruction of farm from the second occupation phase based on preserved wooden remains – Wikipedia

Figure 3.5 Mycenae, the Lion gate – Wikipedia

Figure 3.6 The origins of geometry in Stanley Kubrick's film *2001: A Space Odyssey*

These building practices are not a specific aspect of antiquity but are a characteristic feature of everyday life. They are acquired in order to drive a car or fly a plane. They are handed down over centuries in the fields of agriculture and diet.

Building practices sometimes evolved and spread with great simplicity. An obvious example is rural architecture, based on building practices that could be easily handed down. On other occasions, they required long, arduous practice, as emerges in the main experimental training – carried out directly on site – of the great builders of the past. Over time, it is clear that prototypes were established and building practices were defined which the genius of great architects would find a way of reinventing. This set of rules became consolidated and formed the basis of the material culture of construction in all its various guises, such as the building site, the materials, the planimetric and volumetric view and the scale of the various structural elements. For example, the choice of sites for the construction of many settlements was almost always extremely cautious, eventually achieving stratification that lasted thousands of years. Unfortunately, seismic vulnerability was sometimes overlooked, although this was due to the deeply held conviction that earthquakes were a sign of divine punishment.

Similar considerations apply to the scaling of the various structural elements, which vary from region to region according to the diversity of natural materials. However, builders in antiquity, thanks to centuries of experience, were clearly aware of the lack of homogeneity of materials and their anisotropy (the variations in their physical properties along different directions); these properties affected their use. Lastly, builders would sometimes use models so that the people commissioning the work could make an assessment, but also partly to transfer their view of the structure to the people who had to make it.

3.2. The advent of the science of construction and industrial materials

Right from the earliest days of the scientific revolution, the resistance of materials was a matter of great interest, as demonstrated by the first experiments conducted by Galilei (Figure 3.7). However, the development of mathematical analysis led to the formulation of rational mechanics, which represents the analytical and geometric interpretation of nature in abstract form but which enabled the formulation of key principles, as well as the cardinal equations of statics and dynamics. It made it possible to formulate mathematical models which interpreted natural laws in an increasingly rigorous way. Rational mechanics led to the development of the sciences of engineering and, in particular, the science of construction, which, together with the emergence of industrial materials, created a complete revolution in the world of building from the mid-nineteenth century onwards. Over a brief period, the training of the engineer-architect evolved and developed into an analytical vision which made it possible to shape materials according to mathematical models through unequivocally verifiable calculations (D'Agostino 2008).

New materials available to the building industry included iron and concrete, which were symbiotically linked in the form of reinforced concrete. This triggered a far-reaching revolution after thousands of years, during which the art of construction had created the "emblema civitatis" that had underpinned all cultural views over the millennia (D'Agostino 2015).

Modern architect-engineers have abandoned experimental practice and now train in university lecture halls, learning rational mechanics, science and constructional technique. Their expertise is based on the knowledge of the physical and mathematical theories which

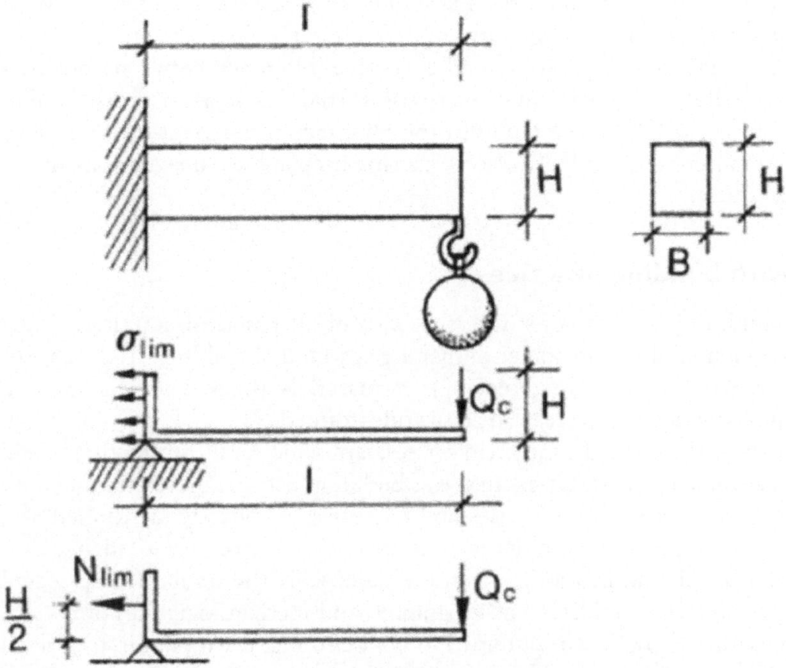

Figure 3.7 Some Galilei experiments: the shelf – Wikipedia

enable the design of modern constructional elements: beams, pillars, frameworks, floors and shell structures. For a lengthy initial period, the design process took place in a fragmented fashion, given the limited effectiveness of calculation tools, but the advent of digital calculation meant that design could develop organically in all its complexity.

3.3. Construction–structure

The significant difference between the ancient art of construction and modern building is highlighted in the construction–structure dichotomy.

Construction is a self-sufficient product designed using an overall spatial view in which the "load-bearing" function is only one of the many functions that preside over the design. It is part of a spatial view, which is elevated in an overall view from the foundations to the roofing. It is a profoundly unitary organism designed to create a single monolithic complex.

The foundations are continuous and almost always direct. The materials are traditional and vary from region to region. The executive techniques, refined over centuries of experience, influence static behaviour. The scaling takes place using building practices that are handed down from generation to generation, and occasionally according to the enlightened intuition of great architects.

The *structure* is a skeleton made of industrially produced materials, designed according to mathematical models that take shape within a technical theory that respects regulations which are legally binding in some countries. It is made in its entirety, and only subsequently is it covered with the elements of industrial buildings, known as *finishes*. The most widespread constructional elements are frameworks, made of beams and pillars, and floors. More recently, in the large complexes of contemporary architecture, structures take on more complex forms. The conceptual, design, material and executive difference between construction and structure underlies the problematic relationship that has marked the structural engineering and constructional conservation of historic heritage throughout the twentieth century. This has led to alterations and distortions, causing irreparable damage to the material history of architecture.

3.4. Designing with building practices

As has been seen, building practices constitute the basis of all practical activities. They represent the canons of material culture aimed at making any kind of building. In the case of architecture, they were structured in the choice of plot, materials, the scale of the building and its planimetric and volumetric features. The basic rules formed part of a fairly widespread heritage for ordinary construction and concerned houses, farms and rural architecture, while knowledge and skills became increasingly exclusive as buildings became more monumental, such as churches, aristocratic residences, castles and fortresses. As already mentioned, the choice of site for the building of ancient cities is quite astonishing, given the prudence that ensured thousands of years of stratification. This was undoubtedly the result of deep-seated knowledge of what would today be classed as orographic and meteorological information. Although houses were initially built without any kind of planning, town planning evolved rapidly and, by the fifth century BC, the Hippodamic grid layout was widely used in Greek cities and can still be observed in the historic centre of Naples.

In the art of architecture, the rules were based on simple modular ratios which became a basic feature of cultural traditions used to create the urban and rural housing fabric. These building practices enabled the creation of towns and cities and the details will be discussed later on in the volume in a series of case studies.

Firstly, it is necessary to observe how the building practices became formalised through lengthy processes and underwent regional variations, linked primarily to locally available materials. The evolution of great monumental prototypes took place over centuries, as is the case for the typological configuration of the classical Greek temple, which could vary in time and space. A highly perceptive essay by Dieter Mertens (Mertens 2015) ends with the quotation of the suggestion about the hierarchical principle of proportions: (1:2) for the main temple, (3:2) for the agora and (4:3) for buildings, dating back to the ideas of Pythagoras (Figure 3.8).

A particularly interesting constructional typology for Roman architecture consisted in *substruction*. Figure 3.9 shows the cross-section of a substructure marked by modular measurements of its various constituent parts. The Baths of Caracalla provide the example closest to hand for L=4 m. Roman substruction, with variations ranging from 3 to 5 metres, found endless applications in Roman architecture, such as the Senatorial palace, the Theatre of Marcellus, the amphitheatres and the substructures that supported the Palatine. The simplicity of this modular approach enabled the rapid construction of substructures in building sites throughout the empire.

The choice of materials depended on the functions and importance of the monument; for example, more resistant material was adopted for the lower floors. A simple calculation using the limit-state method reveals the full stability of a building with these proportions.

Figure 3.8 Agrigento, the Temple of Concordia – Wikipedia

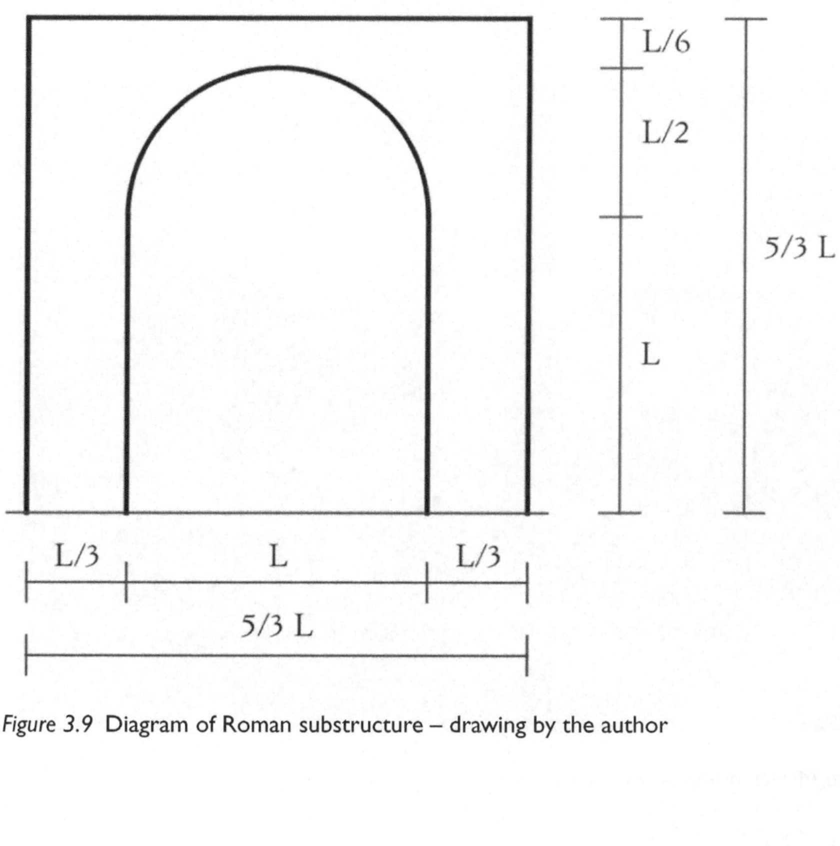

Figure 3.9 Diagram of Roman substructure – drawing by the author

The use of several substructures clearly increased resistance, which is proven by the survival of substructures throughout the empire after two millennia (Conforto, D'Agostino 1995).

Similar hypotheses can be made for basic constructional prototypes found throughout the Mediterranean (Conforto, D'Agostino 1997). A direct derivation of the substructure is the Roman piered bridge. The arched stone bridges seated on grids of tightly packed piles were extraordinarily resistant, as is demonstrated by the large number of bridges all over the former Roman empire that are still extant and often in use. In view of the difficulty of building deeper foundations, Roman architects increased the size of the piles in order to ensure a sufficiently large area of sedimentation; they also moulded the piles in the area just above the water surface, with cutwaters upstream and downstream of the bridge to reduce the effect of the current water. In many cases, such as the Milvian Bridge in Rome, they placed large foundations of slabs on the river bed upstream of the bridge, which helped to regulate the current and prevent the need for a ditch further upstream (Figure 3.10).

In recent decades, the Roman bridge of Porto Torres in Sardinia, without undergoing restoration or alterations, has borne the enormous flow of transport for the adjoining petrochemical plant. In this case, too, structural calculations demonstrated the relatively low stress state despite the massive loads of the heavy-goods vehicles.

A fundamental prototype for Western architecture was the apartment building as it developed during the Roman period. These types of buildings constituted the basis of the architecture of all European cities, and the palace of Augustus or the Basilica Giulia in Rome can be considered illustrious forerunners (Figure 3.11).

Figure 3.10 Rome, Milvian bridge – Wikipedia

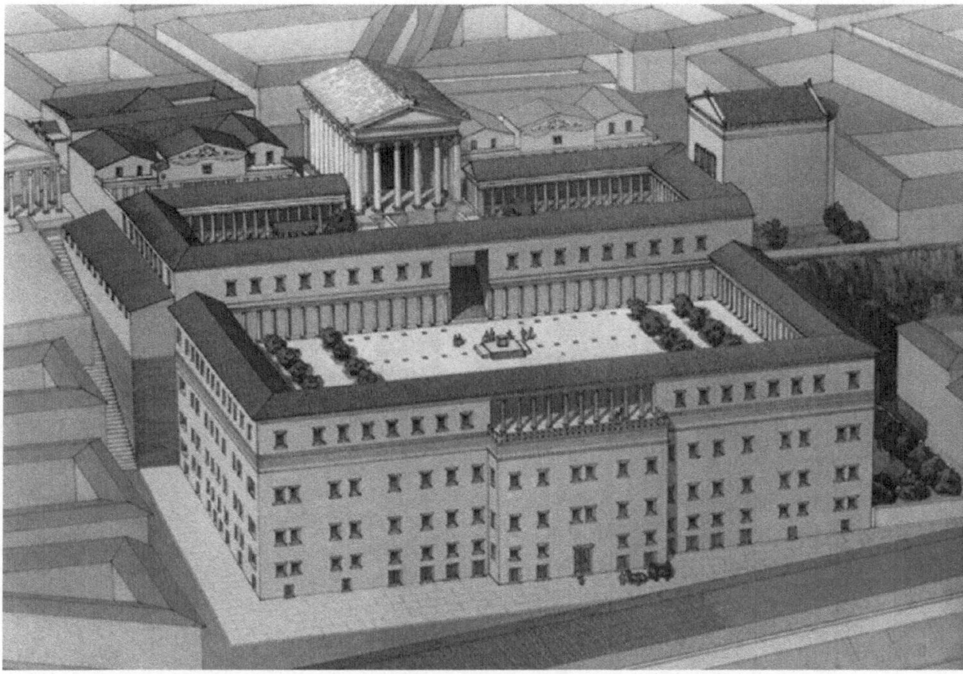

Figure 3.11 Rome, reconstruction of the House of Augustus – Wikipedia

Apartment buildings generally consist of a masonry box made from walls that can be 4–5 storeys above ground level, with a foundation layer of variable depth but which is often 4–5 metres, creating a basement. The material employed is the local stone of the region where the building is situated, or brick. The walls, which are securely joined together to create a spatial volumetry, are of a thickness that varies with the height; while they could measure about 40 centimetres at the top, the foundation walls reach and sometimes exceed a thickness of 1 metre. They are bound together by strong lime mortar. The masonry structure creates gaps between the walls that are usually 4–6 metres, which represent the span of the floors, traditionally made of wood, with framework between the walls and joined to each other. On the lower levels, the floors are often replaced with vaults and staircases and landings are frequently made with depressed vaults. Figure 3.12 provides a schematic diagram of an apartment building.

As well as apartment buildings, Western architecture consists of churches, castles, towers, fortified defensive walls, etc. Specific building practices existed for all these types of buildings, although they are harder to identify; churches, in particular, even within their own liturgical specifications, often vary depending on the styles in vogue. However, all masonry structures share certain key constructional elements: walls, pillars, floors and ceilings, arches, vaults and domes.

The oldest walls were made of *opus isodomum* with worked stone blocks that were not bound together with mortar. The friction between the interlocking blocks ensured stability. In some exceptional cases, such as prehistoric Sardinian towers (*nuraghi*), the *opus isodomum* was made of small stones bound together by clay mortar.

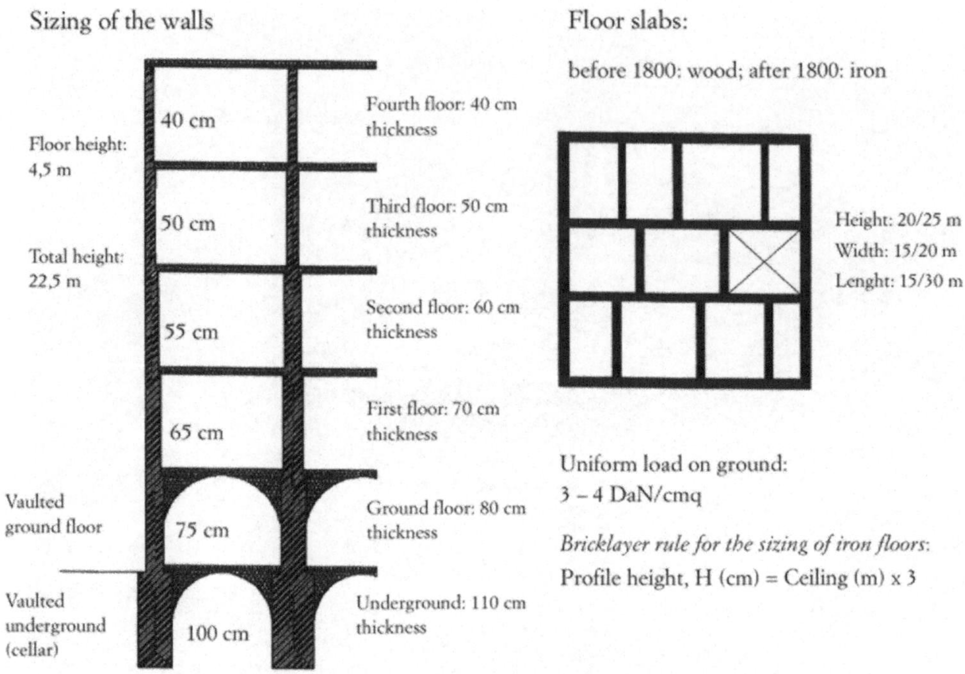

Figure 3.12 Diagram of Neapolitan apartment building – drawing by the author

From the Roman era, masonry walls consisted of two faces between which was a rubble and mortar core known as the *sacco*. The walls were sometimes joined transversely by stone elements known as *diatoni* (headers).

The thickness of masonry walls varied according to their function. For isolated walls, the thickness can be assumed to be equal to a tenth of the height. Specific thicknesses can be found for special types of masonry like defensive walls, towers and castles. Thick masonry is also encountered in the tambours that supported domes. A good example is the tambour of the church of S. Maria del Fiore in Florence, which is 4 metres thick.

Supporting masonry, substructures, buttresses and counterforts made it possible to shape the territory, including roads, canals, embankments, river banks and coastlines.

Pillars can be of considerable size and their thickness can generally be calculated at least as a tenth of the height so that they can be configured as squat solids, safe from the threat of instability. They are sometimes faced with stone blocks, at other times configured with columns or decorated with lesenes (plaster strips). When the arrangement of the upper vaults eliminates the thrusts, they can be extremely slender.

Arches are a constructional feature stemming from the observation of nature, such as vaults and domes. They have existed since the fourth millennium BC in the form of aqueducts or sewerage pipes, made of gradually sloping arches so that they could be made without centring. They make it possible to construct passageways and porticoes and can take the form of round arches, depressed arches, flying buttresses and, occasionally, skew arches.

They can transmit thrusts, provide a contrast between buildings and transfer strains to masonry walls, as in the case of discharging arches.

Supported by centring during their construction, the arches are made of voussoirs (wedge-shaped ashlars) shaped to create a mutual contrast which centres on the keystone. Tie-rods are a device designed to eliminate the thrusts of the arch. Many manuals contain tables for the sizes of arches, but it is worth mentioning the building practices indicated by Leon Battista Alberti in "De Re Aedificatorie" in 1452:

> Finally, whatever the arch used to face the vaults, it should be composed of large blocks of extremely hard stone, similar to that which would have been considered suitable for the piers. The blocks used in the arch should be no thinner than a tenth of the chord. The chords themselves should be no more than six and no less than four times the width of the piers. To hold the wedges together, brass pins and cramps of no inconsiderable strength should be inserted. Then the uppermost part of the arch, known as the spine, should be shaped along the same lines as the others, but have a somewhat thicker head.

These clear guidelines were not always followed, such as in the case of the complex joins between the various voussoirs.

The treatise by Claudel and Laroque (1870) provides the essential measurements of the most important arch bridges made in France during the nineteenth century. For example, the Pont de l'Alma in Paris has three elliptical arch spans, with each span measuring 43 metres, being 8.60 metres high, having a thickness at the keystone of 1.50 metres and featuring abutments of 3.10 metres.

Vaults, which have existed since earliest antiquity, can take many forms: barrel vaults, cross vaults, trough vaults, rib vaults, star vaults, fan vaults, cloister vaults and depressed vaults. They can be made of a solid masonry structure in order to absorb heavy loads or even be constructed as timbrel vaults made of one layer of bricks. Domes or cupolas, a highly prestigious constructional element, can take on different configurations, such as the depressed vault of the Pantheon, the pointed arch of the so-called Temple of Diana at Baiae, the dome of St Peter's in Rome or the rib vault of Brunelleschi at Santa Maria del Fiore in Florence. It is not possible, in the current context, to indicate the building practices of vaults and domes because they depend on many factors, such as the constructional technique, the materials, the techniques designed to lighten the vaults and the different periods of construction, whether Roman, Byzantine or medieval, etc. However, there are classic treatises and more recent works that discuss masonry vaults in historic buildings (Tomasoni 2008).

3.5. Reinterpreting built heritage

During the twentieth century, structural engineering evolved alongside modern architecture in complete contrast to the traditional concept of construction, the theory and practice of which have been considered obsolete.

However, knowledge of the traditional concept of engineering is clearly essential for designing any kind of restoration of historic architecture.

The new structural engineers intervened in a confused, inappropriate manner according to the techniques of consolidation which had no scientific basis and were in direct contrast to the traditional approach to building. In this way, drilling, reinforced concrete, grout injections, plating of reinforced concrete beams and reinforced concrete inserts invaded European historic architecture, causing irreparable damage to the material history of traditional buildings (Carbonara 1981–1984).

Compared to the modern theory of construction, all historic buildings are configured as archaeological constructions since they refer to an extinct culture of design and practice. Under the aegis of the Italian Heritage Ministry (MIBAC) and then the scientific community, a heated debate took place in Italy regarding the inadequacy of consolidation, leading to the theory of conservation, which will be discussed below (Com. Naz. Prev. Sismica 1986, 1989). Unfortunately, consolidation techniques have continued to be widely applied.

Numerous treatises were published between the eighteenth and nineteenth centuries that record the development of the mechanics of masonry. They studied the various elements of construction without an overall vision of monuments, as is clearly shown in the fascinating volume by Becchi and Foce, *Degli archi e delle volte* (2002). Nevertheless, there remains a discrepancy between mechanical interpretation and the ingenuity of ancient architecture, revealed by the authors in their analysis of the wonderful vault of the Hotel de Ville in Arles, designed by Mansart and constructed by Peytret:

> the work demonstrates the excellence of the construction and the idea underpinning it. However, we lack the instruments to understand the reasons for its effectiveness: they are instruments we no longer possess because they belonged to the experience of the master masons who, from generation to generation, had reinvented their own expertise through bold intuitions which were eccentric compared to the leading treatises of the period. (Becchi and Foce, op. cit.)

Unfortunately, given the paucity of available documentation, which consists mainly of architectural surveys that lack measurements of the various constructional elements, it is extremely difficult to conduct research into the building practices through direct study of the monuments. Theoretical research has been carried out that schematises the masonry fabric in its unitary spatial complexity consisting of rigid blocks and panels (Heymam 1997). This approach has recently been described and discussed in great detail in the volume by Mario Como (Como 2010) on the static behaviour of historic masonry buildings. The author suggests in each case the possible static behaviour of various constructional elements, identifying feasible static solutions that justify the static efficiency of the work. Various interpretations will inevitably be possible, and different authors, following differing hypotheses about statics, have demonstrated the functionality and static efficiency of several constructional prototypes. A classic example is the case of staircases with depressed vaults (Roman staircases) which are such a distinctive feature of monumental architecture of southern Italy; two scholars, Baratta (2007) and Como (2010 op. cit.) demonstrate their stability, despite following very different approaches to static behaviour. Como argues:

> To this day the technical literature lacks a unitary, consolidated approach to the analysis of the static behaviour of masonry buildings which can, to some extent, be compared to the approach that exists for buildings made of reinforced concrete or steel.

They were designed in an experimental historical cultural environment which differed profoundly from the mechanical concept. An extremely useful approach for testing, used widely in the past and now completely obsolete, was graphic statics, for the application of which the resistance to the traction of masonry was deemed to be zero. In particular, Mery's method for calculating stone arches is mentioned (D'Agostino 2016).

3.6. The knowledge required for conservation

Unfortunately, over the last 50 or more years, the hallowed traditions of construction no longer form part of the culture of architecture, nor even of engineering. The numerous studies of architecture and buildings made during the eighteenth and nineteenth centuries have been completely ignored by the modern scientific approach, without realising that it remained essential for understanding the ancient concept of construction and informing the conservation of historic heritage, in accordance with material history. The wonderful treatises by Lamberti (1781), Claudel and Laroque (1870), Breyman (1884–85), Belidor (1729) and, in Italy, the volume by Masciari Genoese (1915), the first treatise on seismic engineering on traditional architecture, have been completely ignored, and new generations of technicians and architects have no knowledge of the rich culture of ancient construction. They are completely disorientated with regard to ancient monuments and generally tend to interpret them according to the canons of the science of modern construction (D'Agostino 2006). This has paved the way for the ill-fated practice of consolidation that has led to the uncontrolled development of built heritage, which will be discussed in greater depth later on in this volume.

Recently, however, several key hypotheses have been proven in a major publication entitled *Scientia abscondita :arte e scienza del costruire nelle architetture del passato* (Como et al. 2019):

- The validity of the science of proportion and therefore of building practices;
- The importance of gravity in the stability of masonry structures;
- Modest tensile strength;
- The inadequacy of classic construction science for understanding the concept of construction in antiquity.

What emerges, from a strictly scientific perspective, is the extraordinary skill of ancient builders, their knowledge of "geometry and Pythagorean principles" and, at the same time, the profound dichotomy between the concept of architecture in antiquity and modern architecture.

References

Baratta A., [2007], Sulla statica delle scale in muratura alla "romana", in *Notiziario Ordine degli Ingegneri della Provincia di Napoli*, no. 6, pp. 39–50.

Becchi A., Foce F., [2002], *Degli archi e delle volte*, Venice, Marsilio.

Breymann G.A., [1884–1885], *Allgemeine Bau-Konstruction-LehremitbesonderBeziehung auf das Harbbauwesen*, Salzwassen-Verlag Ed., 2014.

Carbonara C., (ed.), [1981–1984], *Restauro e cemento in architettura*, Rome, AITEC.

Comitato Nazionale per la protezione sismica del Patrimonio Monumentale

Circolare, 1986 *Interventi sul patrimonio monumentale a tipologia specialistica in zone sismiche: raccomandazioni*. Circolare 18 July 1986, no. 1032.

Circolare, 1989 *Direttive per la redazione ed esecuzione di progetti di restauro comprendenti interventi di 'miglioramento' antisismico e 'manutenzione' nei complessi architettonici di valore storico-artistico in zona sismica*, both in La protezione del patrimonio culturale. La questione sismica, R. Ballardini (ed.), Gangemi Ed., Rome.

Claudel J., Laroque, L., [1870], *Pratique de l'art de construire*, Paris, Dunod.

Como M., [2010], *Statica delle costruzioni storiche in muratura*, Rome, Aracne.

Como M., Iori I., Ottoni F., [2019], *Scientia abscondita, arte e scienza del costruire Nelle architetture del passato*, Marsilio Editori, Venice.

Conforto M.L., D'Agostino S., [1995], *Sulla concezione strutturale dell'architettura antica, un caso emblematico: la Sostruzione romana*, Atti XII Congresso Nazionale AIMETA, 3–6 October 1995, Ed. Giannini, vol. II, Tomo 1, pp. 101–106.

Conforto M.L., D'Agostino S., [1997], *Mechanical Resistance and Structural Behaviour of Some Constructive Prototype in the Mediterranean Area*, Proceedings 4th International Symposium on the Conservation of Monuments in the Mediterranean, Rhodes, 6–11 May 1997, vol. 2, pp. 561–570.

D'Agostino S., [2006], *Dalle Regole dell'Arte alla Scienza delle Costruzioni*, Bollettino Restauro Archeologico, Alinea Ed., Florence, 1/2006, pp. 9, 12.

D'Agostino S., [2008], La storia del costruire e la meccanica razionale, in AA.VV. *Mathematical Physics Models and Engineering Sciences*, Naples, Liguori, pp. 175–189.

D'Agostino S., [2015], Between Mechanics and Architecture: The Quest for the Rules of the Art, in Aita D., Pedemonte O., Williams K., (eds), *Masonry Structures: Between Mechanics and Architecture*, Birkauser-Springer International, Basel, pp. 1–19.

D'Agostino S., [2016], La statica grafica e il calcolo delle strutture, in D'Agostino S., (ed.), *Atti del VI Convegno Nazionale di Storia dell'Ingegneria*, Naples, Cuzzolin.

Forest de Belidor B., [1729], *La science des Ingegneurs*, A. Jombert, Paris.

Heyman J., [1997], *The Stone Skeleton: Structural Engineering of Masonry Architecture*, Cambridge University Press, Cambridge.

Lamberti V., [1781], *La static degli edifice*, Forgotten Books, Naples, Nuova Edizione, 2019.

Masciari Genovese F., [1915], *Trattato di costruzioni antisismiche preceduto da un corso di sismologia*, Hoepli Ed., Milan.

Mertens D., [2015], Il tempio classico antico: concetti e progettazione; La ratio dei Greci nel costruire, in *Atti VI Conv. Di Storia dell'Ingegneria*, vol. I, Cuzzolin Ed., Naples.

Tomasoni E., [2008], *Le volte in muratura negli edifici storici, tecniche costruttive e comportamento strutturale*, ARACNE, Rome.

Chapter 4

Heritage conservation

The identity of a country's environment, landscape and territory is a reflection of its history and forms part of the profoundest roots of its people. This is particularly true for European architecture, which has shaped the villages, towns and cities of its countries for centuries. It constitutes an archive of the material history of the various civilisations that have succeeded each other over the millennia. This archive needs to be handed down to future generations as an "emblema civitatis". The conservation of cultural heritage, especially architectural and environmental heritage, is a cornerstone of European culture. As has already been emphasised, contemporary architecture marks a radical departure from the canons of the past and, by creating a fault line, has effectively turned built historical heritage into archaeological heritage, with important consequences for its conservation. It would be interesting to carry out a rigorous analysis of modern architecture and, in particular, urban architecture, though this does not lie within the scope of this book. However, certain aspects are worth examining. Firstly, modern architecture has managed to create a new culture of inhabiting space, ensuring comfortable conditions for most of the population in every building. A carefully structured planimetric and volumetric approach and the development of plant engineering have radically changed approaches to housing. The widespread use of contemporary construction materials and the development of the modern science of construction enabled the miracle of reconstruction throughout Europe following the Second World War. This was accompanied both by the enlargement of historic centres and the creation of large suburbs. Unfortunately, however, the peripheries were designed as settlements to marginalise the less well-off, and were beset by problems related to the integration of different ethnic groups. This created a serious social problem, with disruptive pressures on historic centres which have suffered from temporary overcrowding, or protests such as the *gilet jaune* movement in Paris or in the rough outlying neighbourhoods of Rome. Moreover, the latest architectural trends are revolutionary compared to the canons and modules of traditional architecture, with the abandonment of the concepts of order, symmetry and verticality, as is clearly shown by the works of Frank Gehry (Figure 4.1). All this has interrupted the historical continuity of European architecture and has turned conservation and maintenance into specialist aspects of the art of architecture. A process of cultural maturity is currently underway which, by moving beyond the disastrous criteria of consolidation through reconstruction, tends to extend the theory of restoration, now an entrenched idea in art history, to historic architecture, while the contribution of applied sciences is constantly developing, with the increased importance of archaeometry and the emergence of engineering geared towards cultural heritage (D'Agostino 2017). This is reinforced by international institutions that strive to

DOI: 10.1201/9781003160960-4

Figure 4.1 Bilbao, Guggenheim Museum by architect Frank Gehry – Wikipedia

ensure accurate conservation and planned maintenance. The latest document is the Davos Declaration 2018, "which highlights pathways for politically and strategically promoting the concept of a high-quality Baukultur in Europe. It reminds us that building is culture and creates space for culture" (Conference of Ministers of Culture, 20–22 January 2018, Davos, Switzerland). Due to frequent earthquakes which, since the 1970s, have devastated many regions, and the damage caused by consolidation work, a more appropriate approach to conservation has developed in Italy. The wide-ranging debate between the Italian Ministry of Culture, universities and cultural associations has led to the following proposal for a theory of conservation which, however, is often ignored during restoration work (Conti 1981).

4.1. Conservation theory

The theory is based on a series of core concepts. Firstly the monument should be considered as an archive document of the material history of humanity and thus the subject of historical and scientific research. One instrument for the conservation of monuments is routine maintenance. It is necessary to ensure that, as far as possible, interventions are consistent. Although absolute safety is impossible, it is still crucial to search for the highest levels of safety through interventions that preserve the material integrity of the monument. The theory can be summarised in the following principles:

a) The reliability of traditional architecture;
b) Damage caused by weathering and human intervention;
c) The improvement principle;
d) Respecting material history and its conservation;
e) Operating according to the principle of minimal intervention;
f) Planning and implementing routine maintenance.

These principles will be briefly illustrated below.

The reliability of traditional architecture is demonstrated primarily by its survival over the centuries. Monuments are built as an "emblema civitatis", and their construction is pervaded by all the wealth available, since they convey the message that each civilisation seeks to hand down to subsequent generations. Two factors lie behind the dilapidation of buildings: environmental and human-induced violence. Environmental violence reaches its most serious levels in seismic hazard zones, although there are other causes such as bad storms, flooding, fires etc. Human-induced violence is often worse and consists of wars and vandalism, sometimes also as "damnatio memoriae".

This is accompanied by the usual decay of materials which, in the case of traditional ones, is extremely slow and can be contained through routine maintenance (D'Agostino, Marconi 1987). Humanity has always felt the categorical imperative of history, sensing, on the one hand, its fleeting nature and, on the other hand, the profound sense of continuity of the species to which core values should be handed down. This happens in the case of monuments through the transmission of historical memories.

Weathering, as has already been mentioned, can be countered by carrying out routine maintenance, which should be undertaken by ensuring the greatest possible uniformity of materials and techniques. Unfortunately, new construction materials were used throughout the twentieth century. Besides their lack of homogeneity, the precarious nature of these materials soon emerged. Traditional materials are therefore preferable, applied using techniques which adhere, as far as possible, to ancient or historical techniques. The use of plaster and mortar is a striking example. Roman plaster and mortars have survived for almost 2,000 years and remain virtually intact despite the lack of maintenance, whereas modern plasters have a durability that does not exceed several decades (Figure 4.2). Indeed, the precarious state of so many residential and infrastructural buildings constructed over the last 70 years is well known. Luckily, the industry has understood the need to use materials that are compatible with traditional materials and has produced "historical mortars" with pozzolanic volcanic ash. Patch jobs and pointing are therefore possible, as well as micro-injection grouts between ashlar blocks (Figure 4.3).

On the other hand, the construction industry continuously produces innovative materials whose long-term durability is hard to predict. Regardless of the incessant advertisements, they should only be used in the complex world of monumental architecture with due caution.

Improvement is an intrinsic concept of maintenance. It is extremely similar to the concept of prevention in medicine. Assessing the state of deterioration of a building is necessary for prudent intervention in order to preserve its features. This is compliant with the conservation of historic buildings; attempts were made to "improve" the structure, eliminating the damage it had undergone through targeted interventions that safeguarded static behaviour (D'Agostino 2007).

These interventions consisted not only in underpinning foundations, repairing masonry, reinforcing with tie rods, and restoring roofing, but also in eliminating, in some cases, windows and any form of dissymmetry (Strazzullo 1991).

While improvement is a practice included in planning maintenance, it can be particularly significant in the case of areas prone to earthquakes, where levels of safety comparable with those of new anti-seismic buildings are impossible to achieve.

This is unfeasible without the material identity of the historic building being damaged and destroyed. It is worth emphasising that, since it is extraneous to traditional building

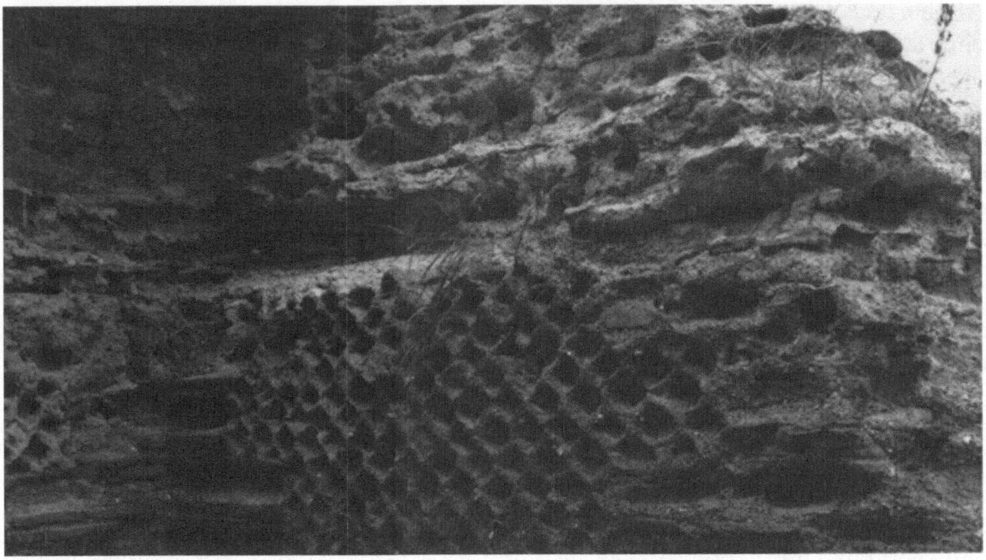

Figure 4.2 Bacoli, Roman plaster and mortar at "Piscina Mirabilis" (Roman cistern) – Photograph by the author

Figure 4.3 Micro-injection grouts between ashlar blocks – Photograph by the author

techniques, modern structural calculation, in all its various aspects, can represent a useful tool for designers. It can be used to reach independent conclusions which suggest suitable interventions for reconstructing, where possible, the original building in a way that would once again render it solid and robust, as has happened throughout its long history (D'Agostino 2002). In this way, designers take on all the responsibility for the intervention, as has always been the case for the great architects and master masons of the past.

Displaying respect for material history and conservation has been a key concept of modern culture since history was conceived of as the "science of mankind" (Bloch 1949), with the emergence of material history. As a monument could now be treated as a document of material history, all the various elements of a building became historical artefacts, just like its planimetric/volumetric dimensions and stylistic characteristics. This can be seen in the study of ancient materials and techniques, in the stratigraphy of masonry and in diagnostic analysis increasingly aimed at the conservation and scientific examination of each aspect of the building. Alongside standard archaeometry, the archaeometry of buildings emerged (D'Agostino 2003), challenging the mistakenly artistic approach whereby a monument is viewed as an artificial structure on which architects can leave their mark (Garzillo 2006).

The principle of minimal intervention is the most complex of the principles set out in conservation theory. Maintenance is an essential form of intervention, but it inevitably involves modifying the building. It is therefore necessary to theorise and implement a balanced approach geared towards the conservation and identity of the monument, as has systematically been done for heritage sites of art-historical or archaeological importance. Unfortunately, however, architectural heritage is used continuously, and adequate safety standards have to be met. Sadly, the current historical era is marked by the requirement of absolute safety, which is unfeasible for historic built heritage. This is a cultural contradiction since, even in European countries with a high level of seismic risk, the number of victims from such catastrophes is undoubtedly lower than the number of victims of common accidents, while there are serious socio-economic problems which could have been largely prevented by careful maintenance. The most reliable form of conservation in Europe has been shown to be projects such as the Monument Watch scheme in the Netherlands, which ensures fairly inexpensive cyclical maintenance. Attentive maintenance means daily care for the most famous monuments, as is clearly shown in Italy by the examples of the Fabbrica di S. Pietro in Rome and the Veneranda Fabbrica di S. Marco in Venice. However, even this type of maintenance has not prevented some dramatic events such as the recent collapse of masonry within the church of S. Croce in Florence with a tragic loss of human life.

Planning and implementing routine maintenance for monumental heritage ought to be a crucial task of the state and institutions responsible for safeguarding heritage. This practice has also been pursued for the prestigious properties of noble families and religious orders. Planned maintenance, which has been acknowledged to be the only correct form of maintenance, is necessary not only for public heritage but also for private heritage and even for contemporary architecture, which often survives in a precarious state.

Just as society places great importance upon its commitment to the prevention of disease, the same should be done to ensure planned maintenance, not just for monumental architecture, but also for ordinary buildings and, to an even greater extent, for the wider network of infrastructural works which are such an important feature of territorial management.

References

Bloch M., [1949], *Apologie pour l'histoire ou métier d'historien*, Paris, A. Colin.

Conti A., [1981], Conservazione, falso, restauro, in *Storia dell'Arte Italiana*, vol. 10, Turin, Einaudi.

D'Agostino S., [2002], Dalle regole dell'arte alla scienza delle costruzioni, in *Atti del Convegno Scienziati-Artisti "Formazione e ruolo degli ingegneri nelle fonti dell'Archivio di Stato e della Facoltà di Ingegneria di Napoli"*, Naples, 5–6 June.

D'Agostino S., [2003], L'Edificio storico: le strutture, in *Atti della Conferenza nazionale di Archeometria del Costruito*, Ravello, 6–7 February.

D'Agostino S., [2007], Il concetto di miglioramento e la sua evoluzione nella valutazione della sicurezza del patrimonio architettonico, in *Atti VI Convegno Nazionale ARCo 'Quale sicurezza per il patrimonio architettonico?*, Mantua, 30 November–2 December 2006, Rome, Nuova Argos, pp. 57–69.

D'Agostino S., (ed.), [2017], *Ingegneria per i Beni Culturali*, Il Mulino, Bologna.

D'Agostino S., Marconi P., [1987], Tecnologie di intervento nel restauro dei beni culturali, in *Atti del I Seminario di Studi del Comitato Nazionale per la Prevenzione del patrimonio Culturale e la Questione Sismica*, Venice, Sez. II, pp. 143–153.

Garzillo E., [2006], *Normopatia, disincanto della Carta di Venezia*, in "Op. cit.", no. 127, Electa Naples.

Strazzullo F., [1991], *Restauri del Duomo di Napoli tra '400 e '800*, Fondazione Corsicato, Naples.

Chapter 5

Traditional and innovative materials

Humans began building structures using materials provided by nature, such as stone, earth and wood, and subsequently iron and brick.

Every single work of ancient and historic architecture was made using these materials, from remote prehistory to the end of the eighteenth century, over a period spanning thousands of years, which witnessed the evolution of civilisations in the various regions of the world.

It was only in the nineteenth century that humankind began to produce new construction materials, and this process was confirmed by the invention of concrete, viewed as the miraculous discovery of a sort of stone with extraordinary chemical, physical and mechanical properties, which could be produced everywhere in the desired quantities and used with great simplicity.

Iron played a key part in the industrial production of materials and was followed by the large-scale production of steel with extraordinary properties, increasingly refined steels, high-strength steels, spring steels and special steels.

From the second half of the twentieth century onwards, the evolution of new materials became so rapid that industry began to design and produce materials of standard uniform qualities.

This led to a rupture with ancient architecture based on the use of traditional materials, which became increasingly confined to decorative functions while they remained the load-bearing material employed throughout antiquity. The following is a brief overview of traditional materials and the main innovative materials that have profoundly influenced the world of architecture and housing.

5.1. Traditional materials

5.1.1 Stone

The use of stone over many thousands of years is linked to the specific regional context and has been a characteristic feature of the oldest archaeological finds from the site of Göbekli Tepe to the civilisations of ancient Egypt and Mexico.

Stone was sometimes been found in situ. A classic example is the construction of the large temples of Paestum, made from travertine, a stone of organic origin that gradually formed over millions of years as a result of the calcification of mud from the river Salso. Other examples of stone found in situ include volcanic materials like the many varieties of tuff, ranging from Neapolitan Yellow Tuff to *peperino*, a stronger grey tuff from Lazio (Latium), which was

DOI: 10.1201/9781003160960-5

used in the construction of major archaeological sites such as Pompeii, Herculaneum and Baiae, as well as constituting the urban fabric of cities such as Rome and Naples.

Sedimentary rocks cover 75 per cent of the Earth's crust and fall into many categories: clastic limestones, limestones, dolomitic limestones, etc., and metamorphic rocks such as marble.

Stone has a high compressive strength, lower tensile strength and reasonable shear strength.

Stone materials are ideally suited to isodomic structures when the blocks are assembled without binders. There is considerable friction between the blocks, which can be increased by the way they are fitted together.

When the blocks are joined together using a binder, they form classic masonry which, due to the immense variety of stone, differs from region to region. Binders have fairly limited tensile strength so that, since the nineteenth century, masonry has been considered a material with poor tensile strength. Ever since antiquity, when sufficient tensile strength was required, ashlars were held together by iron cramps so that the stresses could be distributed evenly. Numerous applications can be found until the nineteenth century, such as the "stone chains" created by Brunelleschi for the dome of S. Maria del Fiore.

From earliest antiquity to the nineteenth century, the whole of European architecture was made of masonry, which was its most distinctive feature.

The conservation of the masonry fabric is an extremely delicate issue which lies at the heart of historic built heritage.

As a result of the introduction of new industrially produced materials, masonry nowadays has an essentially decorative function, and it has become increasingly difficult to find technicians and artisans who possess the necessary expertise to ensure its conservation.

5.1.2 Binders

The first binder ever to be used was earth, in particular clayey soils, which become malleable through the addition of water. In antiquity, buildings were made of unfired clay (*terra cruda*), which is still widely used in some parts of the world. Unfired mud brick requires careful maintenance and is used in entire settlements, such as the ancient city of Bam.

There are plenty of examples that demonstrate the durability of unfired clay. An outstanding example is the reconstruction of a house from the fourth century BC at Terra di Vaglio. The house, which is situated about 1,000 metres a.s.l., has not required maintenance yet is still in excellent condition after over 20 years (Greco 2008). An interesting example of the use of clay as a binder can be found in the construction of Sardinian settlement towers (*nuraghi*) and in their restoration.

A crucial discovery for humankind was lime, produced by heating limestone rocks (calcium carbonate) in large furnaces at 800 °C. It was generally mixed with sand to form the mortar to bond the blocks in masonry building. Roman builders discovered that the mortar formed by mixing lime with *pozzolana* (a pyroclastic sand from volcanoes in Lazio) was much stronger, hardened in water and had hydraulic resistance. With the discovery of lime, in particular lime pozzolan mortars, masonry could make use of various kinds of stone which, using blocks bound with mortar, lime paste, sand and water, could take on various forms and dimensions, as is amply demonstrated by the rich variety of ancient architecture.

Mortar has reasonable compressive strength and modest tensile strength and shear strength. Builders in antiquity were well aware of these aspects, which explains why ancient buildings are generally compact with thick elevations and are therefore very heavy, ensuring stability. Masonry is generally protected from weathering by plaster made using layers of plaster with varying particle size. The durability of this technique is shown by the perfect conservation of Roman buildings built with pozzolanic mortars after two millennia.

Mortar is one of the greatest discoveries of humankind. It has been a constant feature of architecture and houses for millennia through its numerous applications which can be found in masonry, plaster, roofing and floors.

Since the twentieth century, mortar and its components have been the subject of careful scientific research, which now makes it an extremely innovative material with surprising new properties, as will be seen below.

5.1.3 Bricks

The use of terracotta to shape objects and small images is extremely ancient and may date back to the Neolithic, while in Europe, at Paestum, there are beautiful terracotta antefixes and other decorative friezes. Only much later did the Sumerians begin to produce fired bricks, which can be found in Babylonia in the Ishtar Gate, built in the sixth century BC. The use of bricks subsequently spread to Greece and, considerably later, to Italy, where they can be found in the "star column" (*colonna a stella*) in Pompeii, dating to about 120 BC.

Fired clay bricks can be considered the first industrial product in history. They have remained largely unaltered over the millennia and are a typical element of ancient architecture. In the field of building, terracotta was also used for crucial elements such as roof tiles and drainpipes. Roman architecture made widespread use of bricks obtained by heating the Pliocene clays found in many of the hills in central Italy.

Brick was also used to face Roman concrete, made of pozzolanic mortar and shapeless fragments of tuff, to form large walls and pillars. Skilled masons were needed to build the brick walls, while simple slave labour could be used to form and cast the concrete. The compactness of masonry is ensured by the transverse headers which joined the two faces. The use of Roman brick spread throughout the empire and other types are rare.

There are many different sizes of bricks, such as the classic Roman brick, which is fairly thin; and the large sesquipedalian bricks used to build arches. Roman bricks had stamps that indicate the period of their manufacture and their provenance.

Over the centuries, bricks became the key element for building and are a typical architectural feature in many areas of Europe and in Italy, in cities such as Bologna. They were used in different ways, and it is interesting to note their distinctive arrangement in a herringbone pattern in Brunelleschi's dome in Florence. Bricks produced by modern industry in different shapes (perforated brick for facing walls and partitions, hollow blocks and other elements for floors and roofs) still constitute a common element of contemporary building practice.

5.1.4 Wood

Wood, together with stone, marked the dawn of human civilisation. Used primarily as a source of fuel, and as a tool in a wide variety of forms, wood was a crucial material for ancient architecture. The first large Greek temples were made of wood, and were made of masonry only after several centuries. It remains an essential and classic building material in many countries and was crucial for housing, reaching extraordinary levels of craftsmanship, with wonderful frames, elegant furniture and imposing examples of naval carpentry. As long as it is protected and undergoes proper maintenance, wood can last for centuries, as is demonstrated by the large caisson ceilings of the seventeenth and eighteenth century; a striking example is the roof of the ducal palace of Venice, made by the carpenters of the Arsenal and still in place after roughly half a millennium.

There are numerous kinds of wood from the most common to the most elegant, costly ones. It is a light material which is therefore ideal for roofing, and has high levels of resistance despite its marked anisotropy.

Although they are vulnerable to fire, wooden houses have excellent properties: lightness, insulation and resistance to earthquakes.

Due to its lightness, resistance and adaptability, wood has also been used, since antiquity, in half-timbered houses in which it reinforces the masonry.

As will be seen later, wood has been reappraised by the modern building materials industry, which has reconstructed this traditional material, leading to various applications such as plywood and laminated wood.

5.1.5 Iron

The use of iron contained in meteorites is still uncertain, but the Iron Age is usually considered to have begun around 1200 BC, even though the date varies from region to region. It was initially used for tools and weapons and only much later did it begin to find applications in building. A significant example is the use of iron cramps for joining stone ashlars in order to transmit tensile stress. Iron cannot properly be considered to be a building material until the mid-eighteenth century when, with the industrial revolution, it played a crucial role in technological development. During the nineteenth century, iron was such a key element of architecture that it became an integral part of the construction of large buildings and garden greenhouses, and its development went hand-in-hand with the creation of the railway network.

The introduction of iron sections had a crucial effect on building, in particular iron beams, which rapidly replaced the wooden roofs which had been used for thousands of years. The transformation of iron into steel and the rapid evolution of steels were to have a major influence on more recent architecture, from the construction of walls with large spans to the Eiffel Tower, the symbol of the beginning of a new phase of technological progress.

Its fruitful combination with concrete led to the worldwide spread of reinforced concrete, paving the way for a new way of building which represented the first form of globalisation in the aftermath of the Second World War. High-strength steel enabled the construction of prestressed concrete structures, and special steels allowed for a wide variety of industrial purposes. Steel is now the main component of the gigantic, multiform constructions of contemporary architecture, such as Renzo Piano's Shard in London, the Burj Khalifa in Dubai and the forthcoming roofing for the mineral parks of the former ILVA plant in Taranto, a building which is 700 metres long, 254 metres wide and 77 metres tall; it is the largest in Europe.

5.1.6 Innovative materials (Ausiello 2017)

5.1.6.1 Traditional mortars and pozzolanic fillers

By the 1980s, the construction industry, aware of the lack of homogeneity between traditional materials and cement mortars, began producing traditional mortars such as Lafarge and Volteco with pozzolanic properties and a low cement content. Pozzolanic fillers began to become widespread, chiefly fly ash, a by-product of coal combustion in thermoelectric power stations; and silica fume, a by-product of the manufacturing of silicon, which takes the form of an extremely fine powder. More recently, micro-limes and micro-mortars have been developed. The former ensure greater penetration depth, a moderate amount of water and greater compactness. In micro-mortars, the micro-lime particles interact with the micro-particles of pozzolanic fillers. There is a greater affinity with ancient mortars, as well as greater resistance and durability. Micro-mortars prevent the need to pre-dampen surfaces, when of high artistic quality, and can be applied using pumps, by pouring or manually using syringes.

5.1.6.2 Innovative concretes

In the last few decades, concrete manufacturers have introduced significant innovations, launching a wide range of products.

5.1.6.2.1 FIBRE-REINFORCED CONCRETE

Fibres have been used since antiquity as additives. For example, straw and animal hair were used for strengthening unfired clay. The addition of fibres to concrete improves shrinkage and tensile strength, increasing durability. Fibres may be made of various materials: steel, glass or polypropylene. Glass, steel and carbon fibres are known as structural fibres because their elastic modulus is greater than that of concrete, while synthetic fibres such as polypropylene, polyester and nylon, which have lower elastic moduli, are used as fillers; they are also resistant to acidic environments, abrasion and mould.

Fibreglass mesh is particularly effective for wall plaster since it prevents gaps, providing protection against cracking and biological threats.

5.1.6.2.2 NANOREINFORCED AND SELF-COMPACTING CONCRETE

Nanoreinforced concrete uses fibrous reinforcement to achieve unprecedented levels of resistance; carbon nanotubes have extremely elevated mechanical resistance which is about 100 times greater than that of steel. This type of concrete has a highly significant increase in tensile strength, while the compressive strength is much higher right from the very first days of maturity, which makes them particularly self-compacting. It also has high levels of durability, which can be observed from the very first days after the concrete has been cast.

5.1.6.2.3 STRUCTURAL LIGHTWEIGHT CONCRETE

Structural lightweight concrete is marked by the use of a new artificial inert: expanded clay aggregate is added to natural sand as a fine binder. Sand can be substituted by a superfluidifying agent, a viscosity modifier and an aerating agent (foaming agent), which creates microporosity. A low water–cement ratio is essential for mechanical strength, but it is natural in self-compacting mixes due to the presence of the superfluidifying agent.

The mix is produced by soaking the expanded clay for at least 24 hours. Compressive strength is slightly lower, while the shrinkage rate may be greater, although this type of concrete may be self-compacting.

For this type of concrete, the components are all industrial products which ensure constant quality.

5.1.6.2.4 TRANSLUCENT CONCRETE

Mixing concrete with fibre optics, fibreglass or plastic material creates translucent concrete, through which light can penetrate until it reaches the opposite side of the structural element.

As a light conductor, translucent concrete can provide light for interiors by day and illuminate the external environment by night, creating intriguing effects.

A special application was used in the Italian pavilion of the 2010 Shanghai Expo, with panels measuring 100 × 50 cm and 5 cm thick. As well as design projects, it can be used to illuminate cultural heritage sites.

5.1.6.3 Concrete, mortars and self-cleaning, anti-polluting paints

Research has been carried out for some time now to create smart materials that have an active role in producing performance levels that vary over time. The most striking innovation is the use of titanium dioxide nanoparticles, which make cement compounds active since they are sensitive to light. One of the first applications of self-cleaning concrete is the church Dives in Misericordia, designed by Richard Meier and built on the outskirts of Rome between 1999 and 2003.

> In this work, the choice of white is not related just to a linguistic choice … but to a request of technological innovation to regenerate the design of concrete to the nanometric scale, so that it could have access to the field both of sustainable and smart materials.
>
> (Ausiello op. cit.)

Titanium dioxide became a component of mortars and cement paints, creating an active effect. The capacity for self-cleaning was known as the "lotus effect" since the leaves of the lotus, despite being immersed in muddy water, always appear shiny and clean. The formation of a film of carbon dioxide leads to a transformation whereby the surface becomes superhydrophilic about 30 minutes after exposure to the light, a process which ends after about two hours. When solar radiation ceases, the conditions of hydrophobicity are restored.

5.1.6.4 Industrial wood

For several decades, industry has developed procedures for the intensive exploitation of wood, producing plywood and laminated wood. They have the enormous advantage of being high-quality products whose properties are scientifically guaranteed. While plywood is generally made from sheets of wood of various kinds (veneer), laminated wood has become an important structural material due both to its lightness and its reliability. Numerous structures have been made from this material, in particular roofs with large spans for industrial and business complexes, and for archaeological sites.

5.1.6.5 Structural glass

Structural glass is a tempered, or thermally toughened, glass which meets the requirements of mechanical resistance and stability required for structural materials.

A panel consists of two sheets whose thickness varies between 8 and 12 mm, with one or more polyvinyl butyral (PVB) interlayers of a thickness of 1.5 mm.

The most frequent applications are in walls, roofs and beams. It consists of a single column or beam of stratified glass with a moderate span that usually does not exceed five metres. The beams are normally positioned and sometimes have a parabolic shape to follow the bending moment diagram.

A special application is the "Macfarlane" beam, which reaches a span of about 10 metres and is made by assembling several modules.

The panels can also be inserted within a metallic grid with tensile strength. In this case, spans can reach up to 15 metres. The active framework can also lead to the pre-stressing of the panels. This material has evolved rapidly and has numerous structural applications.

The most frequent applications for cultural heritage involve the creation of floors, roofs and staircases for museum displays and in archaeological sites.

5.1.6.6 Composite materials (Balsamo 2017)

Composite materials consist of the combination of two basic materials: a matrix (binder) and a reinforcement (fibre). Their combination defines the chemical and physical properties of the new material.

The continuous phase of the material is known as the matrix, which fixes the fibres to the support, ensures the geometry, protects the surfaces of the fibres, distributes stress between the fibres and ensures shear strength.

The matrix can be organic or inorganic. The former includes fibre-reinforced polymers (FRPs), while the latter includes fibre-reinforced grouts (FRGs) and fibre-reinforced cementitious matrices (FRCMs).

The type of fibres, the quantity and their orientation define the density of the composite material, its breaking strength, shear modulus, compressive strength, ductility, resilience, fatigue damage mechanisms, electrical properties and costs. Structural reinforcements are made exclusively using long fibres, with the possibility of scaling the quantities of reinforcement by overlapping several layers. Long fibres ensure that composite materials have anisotropic properties, so they need to be arranged within the matrix to ensure they are parallel to the direction of the stresses.

Many materials are used for fibres: carbon fibre, aramid fibre, fibreglass, basalt fibre and steel fibre.

Reinforcements in composite materials are usually designed by experts linked to suppliers. In order to work on the design, they have to have a thorough knowledge of the structural organism on which they are working. The type of material will be chosen according to the following criteria: an assessment of the conditions for displaying the reinforcement, which are influenced by environmental conditions and by the finite deformation to be adopted during the design stage: for example, for the nodes of reinforced concrete building frameworks, fibreglass is used for internal nodes while carbon fibre is used for external ones; fabrics can be used with "dry" or "wet" systems; the former is simpler whereas the latter are more complex and need to be installed by specialist technicians.

In the last few decades, FRP systems have been widely used to reduce the seismic vulnerability of masonry structures. They have been used to reinforce the extradoses of masonry vaults, as in the case of the vaults of the Basilica of S. Francesco in Assisi in 1997; for the wrapping of pilasters, walls and ring beams for buildings; and for wrapping columns and anchoring with pultruded bars to reconnect angle irons and masonry ties.

This simple innovative technique, using components manufactured on an industrial scale, leaves doubts regarding its durability which, in the case of historic heritage, should last for centuries.

Conclusions

Traditional materials and the techniques for their implementation form part of a wealth of knowledge stretching back to antiquity. Their history goes back many centuries and ensures efficiency, homogeneity and durability. Safeguarding them requires careful, planned maintenance, which, unfortunately, is rare within the context of architectural heritage management.

The increasingly industrialised business world and artisans are unaccustomed to the use of traditional materials and techniques. This is due to the lack of entrepreneurs working in the construction industry who are interested in the restoration of historic heritage, and to the lack of training in traditional crafts, with the exception of sporadic cases such as stone masons in France. Although a certain level of awareness among conservation organisations has managed to reduce the use of unsuitable materials in large monumental complexes, this is not the case for the constructional fabric of thousands of European towns and cities which reflect the history and culture of different regions. Lastly, industrial products are increasingly designed to have a limited period of effectiveness in order to encourage consumers to replace them. Recently, however, conservation organisations, prompted by the failure of consolidation and by a new technical approach influenced by engineering in the field of cultural heritage, have become much more sensitive to the restoration and conservation of the material values of ancient built heritage.

References

Ausiello G., [2017], Materiali innovativi e Conservazione, in D'Agostino S., (ed.), *Ingegneria per i Beni Culturali*, Il Mulino, Bologna.

Ausiello G., D'Agostino S., [2016], Innovazione e Conservazione: nuovi materiali e realtà virtuali, in *Eresia e Ortodossia del Restauro*, Conv. Scienza e Beni Culturali, Arcadia Ricerche, Bressanone.

Balsamo A., [2017], I materiali compositi FRP per il consolidamento e la mitigazione della vulnerabilità sismica di strutture in muratura, in D'Agostino S., (ed.), *Ingegneria per i Beni Culturali*, Il Mulino, Bologna.

Greco G., [2008], Costruire con la terra cruda: Un esempio dell'antichità, in *Atti del 2° Conv. Naz. Di Storia dell'Ingegneria*, Cuzzolin Ed., Naples.

Chapter 6

Seismic risk from emergency to reconstruction

6.1. Vulnerability of built heritage

Humans have radically modified nature through their constructions and have influenced the landscape in many different parts of the planet. From this perspective, the landscape is an integral part of the character of places and belongs to the built heritage, which is vitally important for establishing its identity. Vulnerability can be defined as susceptibility to damage and concerns multiple aspects that can be briefly summarised as follows: geological and morphological, environmental, hydraulic, structural, seismic, meteoric, thermo-hygrometric and anthropic. These aspects therefore affect all aspects of civil engineering, as well as society itself, with respect to anthropic vulnerability.

6.2. Vulnerability of historic heritage

In general, historic heritage involves buildings dating from remote antiquity to the early phases of contemporary history, which can be roughly dated to the last century. It therefore concerns the history of construction and housing until the beginning of the twentieth century.

A brief summary is provided below of the different types of vulnerability, the effects of which need to be tackled through maintenance and constructive restoration where necessary.

6.3. Vulnerability of modern buildings

In architectural and constructional terms, the term *modern buildings* refers to structures constructed after the introduction of industrial materials, in particular steel and reinforced concrete. They can be subdivided into two broad categories. The first category includes buildings dating from the later nineteenth century to the period of reconstruction following the Second World War, while the second includes buildings constructed in the past 50 years. The difference lies in the fact that buildings in the former category did not have significant seismic sensitivity, whereas it became a standard feature of buildings during the late 1960s and the 1970s. As regards the former, the problems related to the reduction of seismic vulnerability are largely similar to those affecting built heritage, although the situation is somewhat simpler, since it is possible to intervene using the same structural logic with which the building was designed, and the materials are fundamentally similar to existing ones. More recent buildings should have been designed, where necessary, by taking account

DOI: 10.1201/9781003160960-6

of the seismic vulnerability to which they are exposed, and they should only require basic maintenance in the future.

Seismic vulnerability will be discussed in detail below since it is a particularly complex issue in areas of Europe with high levels of seismic risk.

6.4. Earthquakes from antiquity to the present

The history of earthquakes and seismology is a long and complex one. It goes hand in hand with the story of human progress and has sought to verify the occurrence of an earthquake and its intensity, although much work needs to be done before these goals can be fully achieved. A brief overview is provided below of the gradual advances in knowledge of one of the most mysterious forms of natural phenomena, which still causes terrible disasters. In antiquity, earthquakes were always regarded as acts of divine punishment. For example, the Chinese regarded an earthquake as a sign that the gods were displeased with the emperor. This belief was widely shared by peoples throughout the world. There is a beautiful bas relief depicting the sacrifice to the gods following the earthquake of AD 62 in Pompeii, a precursor of the catastrophic eruption of AD 79 (Figure 6.1). In China there are records of earthquakes in antiquity such as the one that took place in 780 BC and an earthquake that happened in Shaanxi province in 1556, causing 830,000 deaths. In AD 132, the Chinese engineer Zhong Heng invented the first seismograph, an ingenious device which, using a pendulum and a small sphere, established the direction of the seismic wave.

The Greeks believed that Poseidon was responsible for earthquakes, but as early as the sixth century BC, Thales realised that there was a natural cause, which he identified as eruptions of boiling water which led the earth to move. In the fourth century BC, Aristotle formulated the pneumatic theory. The *pneuma* was caused by the heat of the Earth or the Sun: if it turns outwards, it produces winds, but if it turns inwards, it causes earthquakes. He classified earthquakes into seven types according to the way they manifest themselves. Aristotle was held in such high regard that this theory remained popular until well into the medieval period and, at a cultural level, considerably later. During the Middle Ages, a theory emerged which maintained there were two types of earthquake – a divine event and a naturally occurring one – but the writings of Albertus Magnus and Thomas Aquinas supported the Aristotelian interpretation, even though Monetti restated the dual interpretation in "De teremoter libri tres", published in 1475. Theories about earthquakes increased during the Renaissance, and Leonardo da Vinci suggested that they originated from underground fires that heated up huge masses of water. One particularly interesting work is the "Treatise on Earthquakes" (*Trattato de diversi terremoti*) written in 1571 by Pirro Ligorio, who had witnessed at first hand the damage caused by the earthquake at Ferrara; he came up with

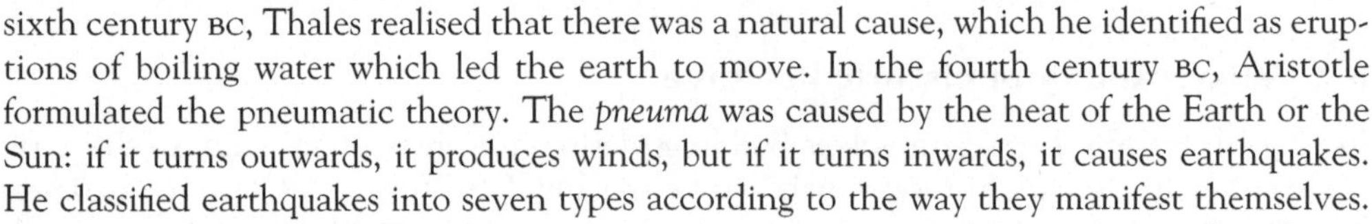

Figure 6.1 The sacrifice to the gods following the earthquake of 62 AD in Pompeii (bas relief in House of Cecilius Giocondus) – Wikipedia

the first design for an earthquake-resistant house and also mentions the volcanic explosion at Pozzuoli (near Naples) in 1538, which led to the formation of the hill known as Monte Nuovo (Figure 6.2). In 1600, Pierre Gassendi put forward a new theory according to which bags of gas accumulated within the Earth's crust and then exploded. In 1626, Niccolò Longobardi wrote a treatise on earthquakes in Chinese.

In 1669, Francesco Travaglini wrote *Physica Disquisitio*, in which he made important observations on the movement of buildings and the waters of the canals of Venice following an earthquake which had struck Dubrovnik and the Republic of Ragusa 800 kilometres away. He studied the propagation of seismic motion, and realised how waves propagate through simple but significant experiments (Guidoboni 2016). In 1779, Pierre Bertholon de Saint Larène put forward a theory based on electricity. Studies intensified after the Lisbon earthquake in 1755, which caused 120,000 casualties with a magnitude of 8.6–9, dealing the city a mortal blow, but only in 1862 did Robert Maillet publish the first photographs of the disastrous earthquake that struck the region of Basilicata in Italy (Mallet 1862). The birth of seismology as a science probably dates back to Eduard Suess, who studied the earthquakes of Austria and published his research in 1873. The International Association of Seismology was founded in 1895.

Seismological studies gradually began to take on a more scientific approach. Given its vulnerability, Italy was at the forefront of this research. An inventory of earthquakes on a planetary scale was compiled at the beginning of the twentieth century, and Richard Dixon Oldham identified various types of seismic waves. In 1902, Giuseppe Mercalli (1850–1914)

Figure 6.2 Earthquake-resistant house by Pirro Ligorio – Il Libro dei diversi terremoti, De Luca. Rome 2005

presented the intensity scale, still in use today, to assess the intensity of earthquakes. It is based on human perception of the earthquake and the damage it causes to the territory. It is divided into 12 degrees. The first is concerned with earthquakes detected only by seismographs and the sixth degree causes damage to houses, while the twelfth means complete destruction.

In 1901, Mario Baratta (1868–1935) published an important study on "Terremoti d'Italia" and the first seismic risk map of Italy. In 1917 he published "La catastrofe sismica calabro-messinese", an important piece of research on the earthquake at Reggio Calabria and Messina in 1908, and towards the end of his life he worked on a revised edition of "Terremoti d'Italia" which was published posthumously in 1937 by the Accademia dei Lincei. Nowadays the intensity of earthquakes is measured using the scale formulated by Charles Richter (1900–1985) as a result of his observations of earthquakes in California. It measures the energy released by the earthquake linked to an indicator of magnitude. The scale has a more precise physical significance and a slightly narrower range than the Mercalli scale. For example, Mercalli scale VII corresponds to magnitude 6.1 on the Richter scale, while Mercalli scale XII corresponds to magnitude 8.1 on the Richter scale. The seismic risk map of the whole world is now extremely advanced and accurate.

The SHARE project, funded by the European Union, collected the available data and carried out a seismic zoning map of Europe. The countries most at risk are Italy, Greece, the Balkans, Bulgaria, Romania and Turkey. The mapping takes account of 30,000 earthquakes with a magnitude greater than or equal to 3.5 on the Richter scale that have occurred since the year AD 1000. There is also a European Archive of Historical Earthquake Data (AHEAD), which cross-checks the data from earthquakes with the data from the 1,100 active faults in Europe over a length of 64,000 km. Lastly, the STREST project seeks to develop a common methodology for assessing the risk for infrastructure, identifying buildings, churches, bridges, streets and pipelines and creating a data bank that is a useful source of information for geologists, seismologists and engineers. In recent decades, intensive historical seismological research has been carried out in order to record earthquakes over the centuries in minute detail, although sometimes, as in Italy, it has revealed not just the country's vulnerability but also the disasters that have taken place in each municipality (Guidoboni, Valensise 2011). A catalogue was published of the earthquakes and tsunamis in the Mediterranean area from the eleventh to the fifteenth century (Guidoboni, Cormastri 2005), as well as earthquakes before the year AD 1000 in Italy and in the Mediterranean (Guidoboni 1989). Studies have also been undertaken in Italy to assess the impact of natural events (Guidoboni, Valensise 2013), and the first manual of historical seismology has also been published (Guidoboni, Ebel 2009). Lastly, extremely important highly specific research has been carried out on individual earthquakes (Guidoboni 2018a and b).

6.5. Emergencies and civil protection

An earthquake can cause casualties, instil panic amongst local inhabitants and inflict damage on buildings and the territory over a fairly wide area. This leads to houses being abandoned, services curtailed and large numbers of homeless people who need to find shelter as rapidly as possible. Over the last few decades, civil protection has become a permanent state institution. Ideally, in each region with a high degree of seismic risk, civil protection would have adequate support structures: firstly, tents, containers, field kitchens and health centres, as well as large quantities of structural components to ensure the safety of buildings, and sufficient numbers of technicians, professionals and engineers who can assess the damage and deal with immediate needs. It might be advisable to set up groups of volunteers for each region ready to be trained in case of an emergency. Lastly, it is essential to remove any rubble that prevents the gradual resumption of normal life. Unfortunately, these conditions are rarely met, and the "emergency" can sometimes drag on for years, disrupting the life of the community and leading to the abandonment of towns and villages that had been inhabited for hundreds of years. It is also worth emphasising land management and badly damaged transport services, pipelines, electricity lines etc., and, lastly, the interruption of industry and business, which can bring a productive system to its knees and undermine the community's economic resources.

Unfortunately, short-term political strategies have been adopted for earthquakes which are structural problems for countries like Italy and Greece. Earthquakes are not dealt with adequately, forcing important areas of the country to experience long periods of disruption which end up having serious effects on the future of territories and the lives of whole generations.

6.6. Constructive reconstruction

Despite repeated political promises, the reconstruction of a territory hit by an earthquake with an intensity of magnitude 6 or more on the Richter scale can prove overwhelming, leading sometimes to the abandonment of the affected settlement or to its reconstruction in a nearby location. In reality, while the emergency leads to a fairly lengthy period of suffering and difficulty, with the rubble remaining in place for many years afterwards, reconstruction is problematic and extremely complex, often having a long-lasting effect on the population. The political decisions taken by the territory following the earthquake are of crucial importance. In many cases, the abandonment of the historic town has proven to be the wrong decision because the new settlement, often built in an anonymous style without any link to local traditions, causes a sense of alienation among the inhabitants, leading to feelings of deep regret.

In general, the state intervenes in reconstruction through grants, both for industrial areas and residential architecture, while the infrastructure is painstakingly restored.

While an earthquake causes terrible loss of human life in the immediate term, it can lead to terrible disruption for the territory and its population in the long term, causing significant damage to its socioeconomic structure.

6.7. Seismic engineering and legislation

During the second half of the twentieth century, seismic engineering underwent significant development through associations and conferences held in Italy, Europe and worldwide as well as extensive scientific research. Regulations had already existed, but major scientific advances were made during this period.

There had been significant regulations in antiquity until the early seventeenth century, when the first houses built with an internal framework of wooden timbers made the building lighter and more elastic. They made use of a system that had already been used by the Romans and can be found in ancient Pompeii. Since then, various regulations have continued to be passed. In Italy, over 30 legislative measures have been passed since the early twentieth century including regulations, decrees and memoranda. The seismic safety regulations and the building codes in Italy and Greece differ from the rest of the European Union; instead of suggested rules, they have the force of law, so they need to be observed closely during the design stage.

The binding regulations are extremely precise and have to be followed by architects and engineers at each design stage.

The problem is exacerbated when it becomes necessary to carry out interventions on historic and archaeological built heritage. As already emphasised, the historic built heritage is an integral part of the material archive of the history of civilisation, and each monument is a living document of this history. As enshrined in Western and European culture in particular, this heritage needs to be preserved as much as possible, as is the case for historical and artistic heritage. Since it is impossible to ensure conformity with the legislation for monuments, let alone archaeological structures, it became clear that the historic built heritage should be excluded from the application of seismic risk regulations; it is to be hoped that the whole of Europe, particularly countries with a high level of seismic risk, will embrace the concept of planned and preventative maintenance, which will reduce many different aspects of risk without the irrational demand for absolute safety.

6.8. Prevention

Engineering should be promoting the need for planned maintenance in the same way that medicine requires preventive medicine. Just as human life comes to an inevitable conclusion, buildings deteriorate over time, and their safety levels become reduced. Planned maintenance ensures environmental protection for the whole territory and an acceptable level of seismic risk for each single building which provides safeguards for earthquakes with an intensity of magnitude 5 and 6 on the Richter scale. Unfortunately, this level of awareness, firstly cultural and subsequently technical and economic, is still extremely vague and hard to apply at a financial level. Plans do exist and would be simple to implement, as will be briefly demonstrated below. Central institutions such as ministries, state and regional governments (such as German *länder*) should be familiar with the details of engineering requirements for cultural heritage and encourage the employment of experts and technicians: architects, engineers, geologists and physicists who advise the institution on the conservation of monuments and archaeological heritage. It is necessary to organise checks and maintenance in the same way as Monument Watch in the Netherlands and Belgium, making sure that each monument has funding for maintenance. This would lead to a significant reduction in dilapidation, enabling periodic overhauls of monuments and more radical intervention only in exceptional circumstances. Similar measures should be adopted for contemporary built heritage, involving local authorities and owners.

Only in this way, when the next inevitable earthquake takes place, will there be drastic reductions in the damage to architecture and the number of building collapses. It is vital to instil investigating magistrates with the idea that engineering is the science of action and the possible, and that technicians should be judged not according to abstract regulations but according to their specific conduct in certain circumstances. Lastly, as has been demonstrated with all human activities from transport to work, magistrates need to be aware that absolute safety does not and cannot exist.

References

Ferrari G., (ed.), [2004–2009], *Viaggio nelle aree del terremoto del 16 dicembre 1857*, Bologna 6 vols. and 3 multimedia DVD ROMs.

Guariniello R., [2016], *Terremoti obblighi e responsabilità. Gli insegnamenti della Cassazione*, Water Kluwer, Milan, Italy.

Guidoboni E., [1989], *I terremoti prima del mille in Italia e nell'area mediterranea*, Istituto Nazionale di Geofisica, Ed. SGA: Storia – Geofisica – ambiente, Bologna.

Guidoboni E., [2016], *Effetti, rimedi e propagazione dei terremoti: nota su un Trattato (1571) e una Dissertazione (1669)*, Atti VI Int. Conf. Di Storia dell'Ingegneria, Cuzzolin, Naples.

Guidoboni E., [2018a], *Florence: The Effects of Earthquakes on the Artistic Heritage: Methods and Historical Sources (15th–20th Centuries)*, Atti VII Int. Conf. di Storia dell'Ingegneria, Cuzzolin Naples.

Guidoboni E., [2018b], *Florence: A First Concise Dossier of Seismic Effect on Artistic Heritage (15th – 20th Centuries)*.

Guidoboni E., Comastri A., [2005], *Catalogue of Earthquakes and Tsunamis in the Mediterranean Area from the 11th to 15th Century*.

Guidoboni E., Ebel J.E., [2009], *Earthquakes and Tsunamis in the Past: A Guide to Techniques in Historical Seismology*, Cambridge University Press, London-New York.

Guidoboni E., Valensise G., [2011], *Il peso economico e sociale dei disastri sismici in Italia negli ultimi 150 anni*, Ist. Naz. di geofisica e Vulcanologia, Bononia University Press.

Guidoboni E., Valensise G., [2013], *L'Italia dei disastri, dati e riflessioni sull'impatto degli eventi naturali 1861–2013*, Bononia University Press.

Mallet R., [1862], *Great Neapolitan Earthquake of 1857: The First Principles of Observational Seismology*, London, Italian translation in G. Ferrari (2004–2009), vol. 2.

MIBAC, [2007], *Linee guida per la Valutazione e riduzione del rischio sismico del patrimonio Culturale*, Gangemi Editore, Rome.

Chapter 7

Constructive conservation

7.1. The historical and aesthetic significance of material

Human artistic activity has always been conducted at both an immaterial and a material level. The ancient arts of poetry, literature, theatre and music have recently been joined by cinema and digital art. Painting, sculpture and architecture create artefacts in which the constituent materials play a crucial role. The artist's choice of which material to use is an indicator of aesthetic intent. While this is clear for painting and sculpture, works of architecture were, until the twentieth century, made of masonry and were conditioned by the materials of a specific region. But even in this case, different kinds of stone and brick have been used by architects to emphasise specific functions and determine the material complexity of each building.

In architecture, the choice of stone, its treatment and the way it is laid out, together with the planimetric and volumetric scheme, affect the unique nature of a building and determine its aesthetic qualities.

While all this seemed quite clear for historical and art works by the mid-eighteenth century (Edward 1777), it was not until the first few decades of the twentieth century, with the theorisation of material history (Les Annales 1929), that the subject of architecture took on a historical, scientific and intrinsically aesthetic importance.

The preservation of material is a categorical imperative both for material history and for the aesthetic identity of the monument. It therefore becomes the subject of scientific research and aesthetic judgement. But material is a living entity. It ages, decays and evolves, and its chemical and physical properties are altered. This is why materials engineering has the task of restricting damage and natural ageing.

To this end, it is necessary to carry out a scientific study of the chemical and physical properties of the materials, avoiding the use of incompatible materials that would alter the intrinsic properties of the original ones. Since an intervention alters the original design concept of the building, it should be designed to be reversible, where possible. An accurate diagnostic study of the existing materials is therefore required to assess the properties and the degree of ageing within a context of multi-faceted skills geared towards an interdisciplinary approach. Recent studies have underlined the need to respect material history (Fancelli 2004) and the compatibility of structural restoration with the architectural concept of a building. Greater cultural awareness is required on the part of the architect and/or engineer, who cannot design structural restoration projects without specific training. The main problems stem both from the diversity of the modern concept of architecture and from the development of new industrial materials which have made the constructive concepts of ancient architecture obsolete.

DOI: 10.1201/9781003160960-7

7.2. Restoration in antiquity

Europe's extraordinary architectural heritage produced by a series of different civilisations has definitely been influenced by the ancient concept of design based on experimentation and the sedimentation of practical rules. This inevitably came about as a result of collapses and damage, which gradually prompted the necessary adjustments.

Over the many centuries of existence of monuments, liturgical and social factors have intervened that have often required lengthy interruptions to building programmes or even radical changes. Interesting examples include the transformation of Romanesque or Gothic churches into the Baroque style, and the radical alteration of castle defences following the introduction of firearms and artillery. Building interventions were carried out over the centuries and have affected the material history of every monument. Unfortunately, not only are the archive records extremely scarce, but also research on these interventions has only been carried out in the last few decades.

A review of case studies of interventions in Italy, from antiquity to the modern era, show the methodologies of restoration and building improvement techniques (*Manuale del consolidamento* [Rocchi 2004]).

The ancient Romans, whose architecture was the precursor for many later styles, had already carried out constructive conservation employing the same techniques that would be used for about 2,000 years and continue to be to this day.

Widely used techniques include indenting, the increase in the thickness of masonry sections, the creation of scarp walls and buttresses, the infilling of apertures and intercolumniation.

During the late imperial period, when Rome's monumental heritage was still on display in all its magnificence, there was immense interest not only in maintenance but also in conservation, and this was taking place in a period when many ancient buildings were being transformed into churches.

The numerous examples of constructive conservation include the following:

The Claudian aqueduct (Aqua Claudia). Ponte S. Pietro, ones of the "bridges" of this aqueduct, had already undergone restoration during the Hadrianic period, when the pillars were reinforced by a wall faced with *opus reticulatum* that was 80 centimetres thick. The conduit of the aqueduct was suspended on high arches which, during the reign of the emperor Septimius Severus, were strengthened by narrowing the aperture and reducing the height by building internal double archivolts (Figure 7.1).

The House of the Charioteers (Casa degli Aurighi) is a large residence of the Hadrianic period which consists of three-storey buildings surrounding a courtyard. During the Antonine dynasty, the pillars were strengthened with a brick wall upon which arches and vaults were placed to create a mezzanine floor (Figure 7.2).

The Palatine Hill was occupied during the Republican period by simple buildings; the emperors, from Tiberius to Domitian, constructed magnificent buildings that altered its stability, so many interventions were designed as supports through terraced sub-structures and buttresses (Figure 7.3).

Roman builders were aware that some buildings were constructed to "last forever", and confirmation is provided in Vitruvius, yet they also realised they would have to deal with recurrent seismic events, as Pliny the Elder stated: "The city of Rome never experienced a shock, which was not the forerunner of some great calamity". Various measures were therefore taken to improve anti-seismic resistance.

Figure 7.1 The inner double archivolts in the Claudian aqueduct (Aqua Claudia) – Rocchi: Manuale del Consolidamento 1994

Figure 7.2 House of the Charioteers (Casa degli Aurighi), Ostia – Rocchi: Manuale del Consolidamento

Given their limited height, many buildings in Pompeii lacked indenting in the wall corners, although this rule was only systematically applied after the earthquake in AD 62. Other measures were widely used: buttresses and counterforts, arches joining buildings, travertine indenting joining the wall and the buttress, intradosses, arches and lintels with indentation, the use of pins and cramps to absorb tensile stress, chains and other iron elements, the thickening of pillars and columns, and of colonnades with infilling or blind arches (Cairoli

Figure 7.3 Domus Tiberiana on the Palatine Hill, Rome – Rocchi: Manuale del Consolidamento

Giuliani [2011]). This awareness and these measures have been handed down over the centuries until the beginning of the twentieth century and are substantially the same as those indicated for masonry buildings in modern anti-seismic engineering treatises (Masciari Genovese [1915], Mastrodicasa [1948]).

Theodoric the Great (493–526) is renowned for recommending the restoration of many monuments, while Cassiodorus stated: "Though our intention certainly is to construct new buildings, we are more deeply concerned to preserve old ones, since we can obtain equal glory from innovation and preservation" (Viscogliosi 2004).

The same types of intervention used in the Roman period were adopted during the Middle Ages. In some cases, monuments were reconfigured through demolition and restoration.

Work on the *Garisenda Tower* in Bologna was begun in 1110. Between 1351 and 1360, it was lowered by 12 m to avert the risk of collapse following subsidence which caused the tower to lean, reaching 3.22 m of overhang in a north-easterly direction. In 1399, an earthquake led to the collapse of the belfry; a fire later destroyed all the wooden parts surrounding the tower. Following the fire, the fear of an imminent collapse led to extensive work being carried out to strengthen the lower part through a brick veneer which transformed the interior from a square to a circular section, as well as the reconstruction of the belfry, the restoration of the staircase, etc. It is interesting to note that the tower is built on a base of about 500 m^3 weighing about 1,000 tons, with layers of pebbles bound with lime mortar. The area of clay sediment was consolidated by driving 2-metre-long wooden piles into the ground (Figure 7.4).

The *Dome of Pisa Cathedral* was probably constructed between 1090 and 1100. The elliptical dome with raised arches was designed to counteract the thrust on the rectangular impost. It was built using a basket-weave bond to create a light shell. It probably displayed

Figure 7.4 Bologna, Garisenda Tower – Wikipedia

crack patterns, so a complex intervention was carried out at the beginning of the following century. A series of openings below the dome were bricked up, and a continuous infill wall was constructed with round arches. It was probably strengthened by chains, and its structure and size are unknown. A series of loggias with spires were built above the reinforcement masonry (Sanpaolesi 1959; 1975) (Figure 7.5).

During the Middle Ages, an important element for reinforcing stability was the flying buttress designed to counteract thrusts, as can be seen in the church of *S. Giovanni in Zoccoli in Viterbo* (Figure 7.7).

The arch was also frequently used to counterbalance thrusts between buildings, as can be seen in many Italian towns and cities such as Siena, Perugia. It is an appropriate anti-seismic measure since it manages to disperse the seismic wave over a vast number of buildings.

Structural reinforcement methods remained largely unchanged for centuries.

In Bologna, the brick columns of the porticoes which displayed cracks caused by compression were reinforced with bare stone pillars, while a new arch has been added to the intrados of the existing one.

A similar intervention can also be found in the church of *S. Stefano Rotondo* in Rome; in this case, the pillar incorporates the column, leaving only the front surface and the capital visible.

The church of *S. Eligio degli Orefici* in Rome is an emblematic monument that underwent restoration work immediately after its construction which, for various reasons, continued from 1516 to 1551.

Figure 7.5 Pisa, the Dome of the Cathedral – Wikipedia

Figure 7.6 Tower of Pisa – Wikipedia

Figure 7.7 Viterbo, Church of S. Giovanni in Zoccoli – Wikipedia

After a few years, the foundations were found to be subsiding, so that it was necessary to lay a sub-foundation in the north-western zone, carry out interventions on the masonry, eliminate the lantern tower and restore the lantern. Subsequently, the sub-foundation was enlarged and chains were applied.

Further damage and new collapses took place towards the end of the century, probably caused by flooding of the river Tiber. New work was carried out on the foundation with relining in brick and intervention on the masonry and the dome with a reinforcing rib, as well interventions on the tambour and the lantern (Figure 7.8).

St Mark's Basilica in Venice, whose domes above the vaults decorated with mosaic have a wooden structure, underwent major reconstruction work in the early seventeenth century, carried out by Francesco Sansovino. The damage to the building had numerous causes: the ageing of the materials, especially the wooden elements; a fire; and the probable subsidence of the foundations. The intervention was extremely complex and the building was supported by scaffolding. The interventions probably included work on the sub-foundation, spandrel walls and reinforcement of the masonry; dismantling and reassembly of the wooden domes; and the installation of a chain in the central dome. Further interventions would be undertaken in the second half of the nineteenth century (Figure 7.9).

The *Basilica of S. Domenico* in Perugia has had a particularly complex history. The church, on which construction work began in the early fourteenth century, was the largest Italian Gothic church with lateral naves of the same height as the central one, a typical layout of such buildings in central and northern Europe. In 1611, work was carried out to restore

Figure 7.8 Rome, Church of S. Egidio degli Orefici – Wikipedia

Figure 7.9 Venice, St Mark's Basilica – Wikipedia

the roofing by modifying the structural layout which placed the thrusts on the imposts; the heavy spandrels that verticalised the thrusts were also eliminated. An initial collapse took place in 1614, followed by a further collapse of all the central pillars, leaving only the external structure with the buttresses still standing. The church was later completely restructured by Maderno, who transformed it into a Baroque church (Figure 7.10).

The history of *Salerno Cathedral* is unusual. The cathedral was built in the eleventh century and has a modular layout. Indeed, the module is the width of the transept. The length is five modules while the width is two modules, the height is one and a half modules, the depth of the apse is half a module, its width is one module, the width of the lateral apses is half a module and their depth is a quarter of a module.

The initial signs of damage emerged during the fifteenth century and led to the first intervention: two buttresses were built along the outer wall of the right-hand nave and, in the nave, several arches were constructed that led from the outer wall to the pillars on the right in the central nave. These pillars were made by incorporating the pre-existing columns.

Following the earthquake of 1688, new work was carried out, beginning in 1691. Work was undertaken on the sub-foundation with arches and pillars; additional arches were created in elevation, as well as semi-pillars to support the façade. After an interruption lasting 6 years, work resumed, transforming all the original columns into pillars and joining them to the arches. The external outer walls were also demolished to create a series of chapels on each side. The chapels were founded autonomously and constitute a series of lateral buttresses (Figure 7.11).

Figure 7.10 Perugia, Basilica di San Domenico – Wikipedia

Figure 7.11 Salerno, the Cathedral – Wikipedia

During the sixteenth century, *Palazzo Sanseverino* in Naples was transformed into the current *Church of Gesù*, and Guglielmelli built an elegant dome for it in 1691 following the earthquake of 1688 (Figure 7.12). During the eighteenth century, cracks appeared in the dome and, in 1769, Ferdinando Fuga proposed a major intervention with the creation of counterpillars and counterarches in the central part of the church. This proposal, which would have jeopardised its architectural appearance, was supported by a commission with the exclusion of the architect Gioffredo, who was convinced that the problems stemmed from the subsidence of a central pillar supporting the dome. Although it began, the work was suspended as a result of the opinion of Vincenzo Lamberti, who demonstrated the pointlessness of the intervention proposed by Fuga and denounced the presence of a cistern near the large pillar which had already indicated by Gioffredo.

Unfortunately, Lamberti's opinion was criticised by Luigi Vanvitelli who, however, did not come up with any practical suggestions for intervention, even though he denounced the perilous state of the building. In the midst of these contradictory proposals, the decision was taken in 1774 to demolish the dome, a major blow to the monumental heritage of Naples.

The previous examples highlight the techniques of constructive conservation used in antiquity with examples chosen from Italy, but they are similar to other initiatives carried out throughout Europe. These techniques remained practically unchanged until the advent of new industrial materials. It is also worth mentioning the intervention by Vanvitelli on the dome of St Peter's in Rome, the construction of the buttress for the Colosseum by Stern and the restoration by Stern Valadier of the Arch of Titus on the Palatine Hill.

Figure 7.12 Naples, Church of Gesù Nuovo – Wikipedia

7.3. Constructive conservation in the nineteenth century: "*cementificazione*"

With the spread of industrial materials and, in particular, reinforced concrete, ancient building technology rapidly became obsolete. The new material was considered a panacea, and without any respect for the ancient concept of building and material history, interventions began to be undertaken on built heritage using reinforced concrete. Thus pillars, walls, columns, ceilings and trusses began to be made of reinforced concrete and iron, and ceilings made of hollow-core concrete slabs replaced many wooden roofs that had lasted hundreds of years. An extensive and disruptive application of this type was made by the Greek engineer Balanos on the Acropolis of Athens in the 1930s.

As well as new constructional elements, masonry was filled with partition walls made of reinforced concrete and wire mesh nailed to the masonry, with slabs of just a few centimetres from the extrados sometimes joined to vaults with bars of a small diameter and mesh. The use of steel mesh–reinforced concrete made with perforated masonry and injections of cement-based grout, often involving the insertion of metal bars (Figure 7.13), became widespread. This gut renovation (*cuci e scuci*) was used to "stitch together" the cracks and reinforce arches, columns and pillars (Figure 7.14).

These techniques, which rapidly spread without any scientific support as a kind of new precept, were based on an approximate evaluation of their intensity: four injections per square metre were prescribed for small cracks, while six injections per square metre were prescribed for larger cracks and more extensive damage. In Italy these techniques began to

Figure 7.13 The insertion of metal bars in steel mesh reinforcement – Photograph by the author

be systematically applied following the earthquake in Friuli in 1976 through the widespread dissemination of a lengthy report, and a volume was subsequently published with the lofty-sounding title *Restauro statico dei monumenti* (Lizzi 1981).

Unfortunately, this deplorable technique quickly gained popularity due to the utter simplicity of its design and because it proved extremely remunerative for businesses. The technique won widespread acclaim in Italy at a conference held in Rome organised by the Italian association of concrete manufacturers (AITEC), "Il cemento nel restauro" (The use of concrete in restoration), which led to the cementification (*cementificazione*: the widespread use of reinforced concrete structures to consolidate historic buildings and monuments) of hundreds of Italy's most important monuments (Carbonara 1981–1983).

It was not until much later, due to the serious danger posed by the corrosion of metal bars, that stainless steel, and subsequently fibreglass, was occasionally used instead.

Once again in Italy, there were outcries following the earthquake that struck Irpinia-Basilicata in 1980–1981. There was now growing awareness that the precious archive of material history, the ancient architectural fabric which had evolved over thousands of years of the history of architecture and construction, was being destroyed through an irrational and completely irreversible alteration.

Numerous studies on the reliability of the injection of cement-based grouts were published, leading to significant technological modifications, but unfortunately this research focused on the structural efficiency, neglecting yet again the inevitable alteration of the material history of the buildings involved.

Figure 7.14 Mesh with reinforcing rods – Lizzi: Restauro Statico dei Monumenti/The Static Restoration of Monuments, Sagep editrice, Genoa 1982

Some simplified examples of reinforcements *reticolo cementato*, the widespread use of cement injections and reinforced stitching, are provided below, followed by some classic examples of famous monuments. The data is taken from Lizzi (op. cit.) (Figures 7.15 and 7.16).

In addition to the above discussion and recorded examples, several further considerations are necessary. The stratigraphy of the land of historical sites and, in particular, of monumental architecture, goes back thousands of years and is well known. For example, archaeological excavations have recently begun in the ancient site of Pompeii to study the Samnite settlement, and most churches were built on ancient cult sites, etc. It is obvious therefore that the sub-foundation of a monument with piles or micropiles inevitably causes the complete destruction and cementification of any archaeological layers that are present.

On the other hand, responsible intervention on the sub-foundations of historic buildings with traditional techniques leads to the discovery of important archaeological finds which enrich the history of the site. This has happened with the recent intervention on the Leaning Tower of Pisa with the discovery of ancient necropolises and the ruins of ancient Roman settlements; the conservative restoration of the Rocca dei Rettori in Benevento with the discovery of a cistern and part of a Roman aqueduct, as well as a series of necropolises; not to mention the magnificent fresco found beneath the church of S. Maria Assunta at Positano (Figure 7.17).

Since the mid-twentieth century, historical research has produced major contributions to material history. In particular, specialists in medieval art history have extended stratigraphic

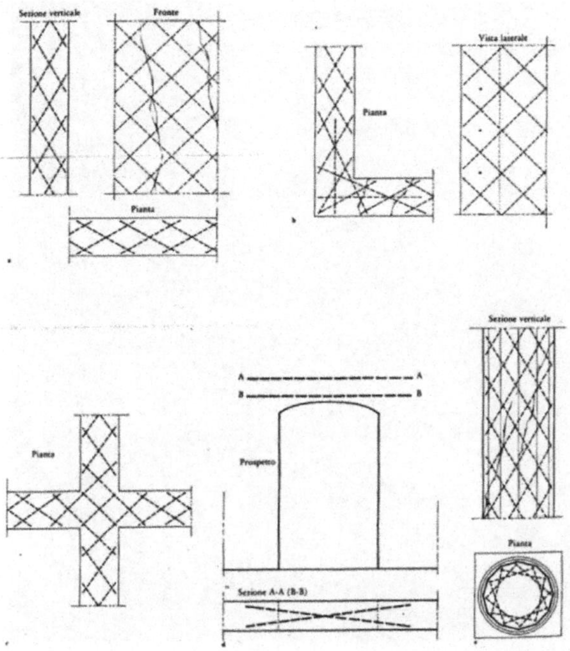

Figure 7.15 Simplified examples of reinforced concrete grid – Lizzi: Restauro Statico dei Monumenti

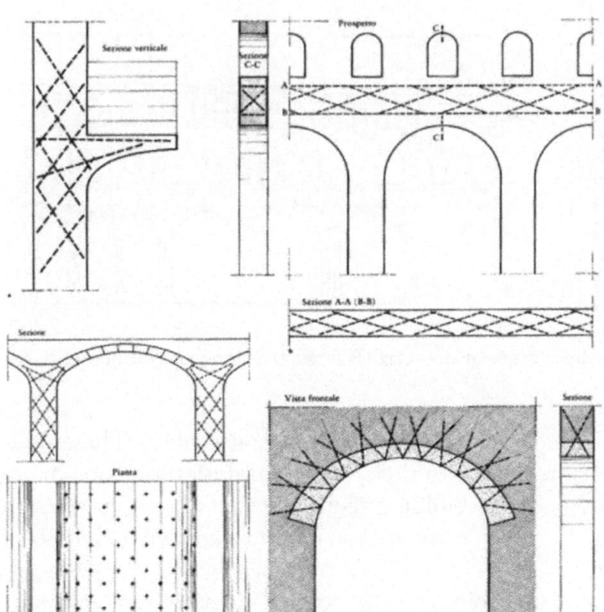

Figure 7.16 Simplified examples of reinforced concrete grid – Lizzi: Restauro Statico dei Monumenti

Figure 7.17 Positano, magnificent fresco in the church of S. Maria Assunta – Photograph by the author

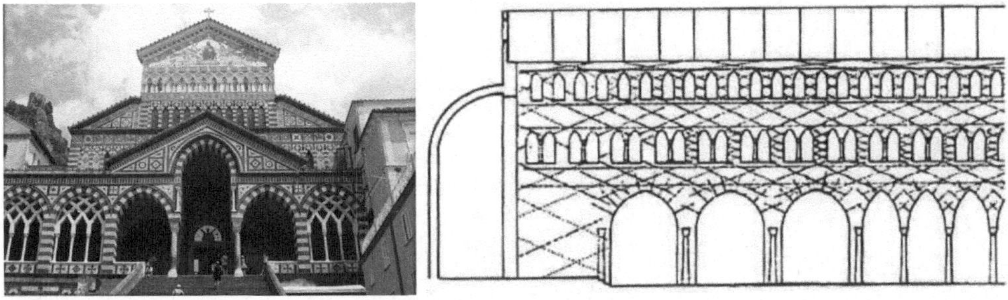

Figure 7.18 Amalfi, Duomo and reinforced concrete grid. – Lizzi: Restauro Statico dei Monumenti

research, already an integral part of archaeology, to the ancient masonry fabric. These studies, which have yielded significant results related to the scale and diffusion of stone materials, their provenance and the analysis of the building techniques employed, have been obliterated by the random and widespread use of cement-based mortars and even more so by cement grout injection.

At a methodological level, the attempt to obliterate the original construction technique by trying to reinterpret the ancient monument with structural inserts belonging to modern construction techniques is particularly serious. This is clear in the case of the Duomo of Amalfi (Figure 7.18) and the intervention on the pediment of the Temple of Athena at

Paestum, where the restoration project involved the use of metal frames which, interpreting the ancient architrave as a continuous beam, with the perforation and reinforcement of the columns, reconfigured the ancient colonnade as a framework. In this latter case, the project had a purely theoretical significance, since the firm carrying out the work inserted metal cramps within the lintel using intersecting diagonals that contradicted the impossible project (Figure 7.19).

While perforations alter and break up the masonry fabric, in the case of seismic activity, normally preceded or followed by a long earthquake swarm, a masonry complex injected and reinforced with bars can be affected by a widespread series of micro-cracks, which undermine the structural efficiency and can lead to collapse.

Lastly, it has been demonstrated that the breathability of masonry restored using concrete is affected with an increase in problems related to damp, which encourages the formation of condensation and the deterioration of the microclimatic conditions (D'Agostino, D'Ambrosio, Riccio 2004).

Unfortunately, consolidation has become a widespread feature throughout Europe, where monuments have been ravaged by intensive cementification (*cementificazione*).

However, it should be underlined that the building industry has managed to tackle the problem much more efficiently than the professional and academic world. It has produced numerous types of so-called historical mortars that seek to reproduce the original ancient pozzolan mortar; moreover, it has fine-tuned systems of micro-injections which do not alter the masonry fabric but fill the existing empty spaces (Figure 7.20).

Lastly, composite material fibres have rapidly become widespread. These fibres, which display significant resistance to tensile stress, can easily be applied to external surfaces

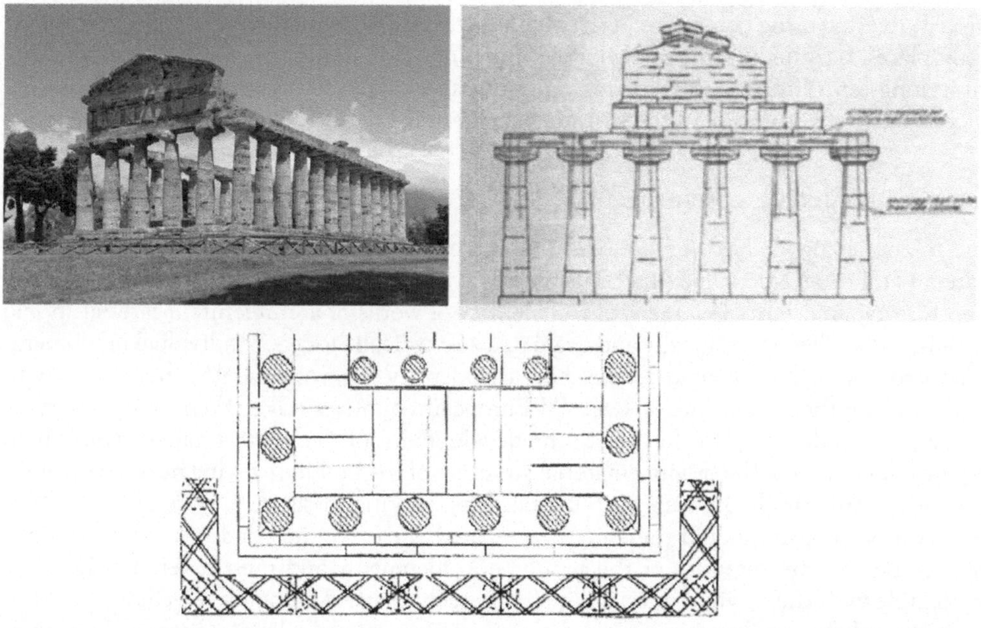

Figure 7.19 Paestum, Temple of Athena: (b) reinforced concrete grid (c) planimetric diagram – Cipriani, Avagliano: Il restauro dei Templi di Poseidonia

Figure 7.20 Micro-injection grouts between ashlar blocks – Photograph by the author

through riveting using composite materials. This technique, which has definitively replaced metal plates, has the advantage of not causing profound alterations to the masonry fabric, but its long-term effectiveness remains doubtful, especially in terms of the centuries required by conservative restoration of the historic built heritage.

7.4. Constructive conservation

Due to the acknowledged importance of material history and the close links between a material and its artistic value, materials have become, both in terms of their substance and their history, an essential element of the identity of works of architecture. Material should therefore be studied scientifically and evaluated aesthetically for each individual monument. In Italy this has led to a heated debate about the identity of a work and its conservation with respect to the theory outlined above. All European countries have taken part, to varying degrees, in this debate. This has led to the development of *constructive conservation*, which requires work upon of the original material structure of the building, giving new responsibilities both to structural engineers and to architects working in the field of restoration. New conservation projects should therefore be prepared with a critical and historical approach that focuses on the integrity of the monument-document, and uses materials which are compatible with the original ones. They should seek to avoid being invasive and should be reversible, where possible, identifying the canons of programmed maintenance. The structural regulations should also be applied with great caution, both during day-to-day operations and in the event of an earthquake.

Several suggestions (D'Agostino et al. 2009) are summarised below. Designed for archaeological structures, they should also be applied to the constructive conservation of any listed historical building, for the reasons discussed above. They constitute useful guidelines for preparing constructive conservation projects for historical heritage.

7.4.1 Foundations

The foundations of a monument are often laid out on an area of archaeological land or fundamentally stable terrain. Unless the area has been affected by landslides, earthquakes or flooding, the interventions for foundations are generally very small-scale. They should be based on exhaustive knowledge of existing foundations as well as the geotechnical features of the terrain concerned. It will be necessary to ensure the greatest degree of uniformity in the support conditions, restoring and reinforcing the horizontal connections or, where necessary, using underpinning, a technique that makes it possible to investigate the archaeological layers. The use of piles or micropiles should be excluded since they would cover the land with concrete, destroying archaeological layers and altering the original relationship between foundation and construction.

7.4.2 Masonry

Historic masonry constitutes a precious and complex piece of evidence for material history in general and for construction techniques in particular. Its integrity needs to be safeguarded more than ever in the contemporary world, where the technological development of applied sciences enables investigations that transform masonry into a historical document of great value. The masonry makes it possible to find out the nature and provenance of the materials employed and the mechanical, chemical and physical characteristics that determine their structural behaviour. The geometric dimensions of the masonry, the analysis of the foundations and, where possible, the study of the building itself provide precious evidence for the ground rules and construction techniques of a specific territory during a specific historical period. Further contributions may come from stratigraphic analysis, and dating methods of inorganic materials, currently undergoing advanced research, should soon bring the analysis of masonry into the foreground of the historical knowledge of any building, defining both the origins and the subsequent interventions. The definition and analysis of the crack patterns and the various forms of dilapidation, when analysed with scientific methods supported by modern diagnostics, can shed new light on the recent evolution of the structure, as well as constituting key elements for the definition of the most appropriate conservative intervention.

During the design phase of improvement intervention, it will be necessary to use materials whose physical and chemical characteristics are as compatible as possible with the original materials. In particular, it is important to use stone materials, bricks and binding agents that are as similar as possible to the ones in the building, while cement-based mortars and inserts in reinforced concrete should be completely avoided.

Measures will be taken according to the specific case: localised repairs and the restoration of damaged parts by crack stitching and, where necessary, through patching (*scuci-cuci*); the elimination of alterations made to the building in subsequent phases and considered to be inappropriate during the restoration project; and indispensable additions required to ensure the stability of the building.

It is essential that the wall undergoing the intervention restores the complete homogeneity of the masonry fabric, uniformity in terms of resistance and continuity in terms of rigidity; where it is lacking, the appropriate structural connection should be created through the re-bonding of the masonry. It is necessary to avoid perforating the masonry fabric with or without the insertion of bars made of steel, fibreglass or other materials, due to their invasiveness of the ancient masonry fabric, their clearly irreversible nature, the legitimate doubts about their durability and lastly, their unknown effectiveness, especially in the event of earthquakes.

In the case of masonry with particularly poor-quality mechanical characteristics, or in the case of micro-cracks and voids in the masonry structure, it will be possible to use more modern techniques using non-cement-based binding agents to try out as a precautionary measure in terms of feasibility and effectiveness, with targeted injections through mortar joints and any eventual cavities. Reinforcement using the plating of reinforced concrete or concrete, and the use of cement-based plasters, should be excluded. These types of intervention destroy the ancient masonry fabric and lead to extensive damp, which prevents the walls from breathing.

Frescoed walls or walls with mosaic both need to be treated with the utmost delicacy. If the fresco only concerns the facing of one side of the masonry, then it is necessary to intervene with due caution on the facing of the other side. However, these interventions should always be carried out by specialists in the specific field.

7.4.3 Pillars and columns

Firstly, it should be emphasised that pillars and columns are generally designed to support vertical loads with modest eccentricity; any action that disturbs this static behaviour may have negative effects on the resistance and stability. Actions of this nature are mainly due to the thrusts from arches, vaults and roofs; in other words, they are caused by dynamic events such as earthquakes and wind action.

To reinforce pillars and columns, the following measures should be adopted: to restore, where necessary, the original resistance under normal stress using solutions such as hoops, iron dowels or supplementary masonry; to eliminate, or at least constrain, the horizontal thrusts using measures such as attaching chains to arches, vaults, roofs and building or restoring buttresses; and to restore the links designed to transmit horizontal actions to more rigid, resistant elements.

Interventions designed to provide columns and pillars with flexural and tensile strength, such as reinforced perforations, pre-stressing and the insertion of metal liners, should be avoided. Indeed, these interventions, besides being invasive and irreversible, are rarely very effective and are often harmful since they drastically modify the behaviour of the overall structure. The dangers in the event of earthquakes have recently been demonstrated (Como 2016).

Wrapping using carbon fibre or similar material, like any other kind of cladding which destroys the structural elements and modifies the structural behaviour, should also be restricted to a minimum.

Non-vertical situations should be analysed carefully, identifying cause and effect and evaluating whether it is opportune to correct or maintain them.

Where applicable, the observations made for the previous section on walls remain valid.

7.4.4 Chains and tie rods

The centuries-old custom of inserting "chains" in masonry buildings has been re-evaluated and should be used systematically, where appropriate, especially in interventions in seismic zones.

Preference should be given to chains made of stainless-steel bars with "bolted end-plates" designed to distribute the force applied by the chain over large areas of masonry; these end-plates should ideally be external to the wall. Alternatively, it is possible to use bars made of fibreglass or other innovative materials whose mechanical and chemical-physical properties, as well as their durability, are proven. Chains are intended to ensure the "box-like behaviour" of buildings, namely their capacity to function structurally as a single spatial organism. Chains should therefore be mainly positioned close to load-bearing walls on each level, preferring solutions using a double chain alongside the walls themselves. In the case of external walls, a single chain should normally be installed on the inside.

> In cases where it is indispensable to pierce the wall longitudinally (to be avoided where possible), it is preferable to use uninjected sheathed tie rods, which make the intervention reversible, enable stretching to resume and prevent anomalous or harmful stresses.

The chains should normally be horizontal. In some cases, it is worth evaluating whether to use vertical or sloping tie-beams.

Whatever the case, the use of tie-beams should always be checked to ensure that the actions caused to the masonry are supported by them with ample margins of safety, even during the transitory phase of installing them.

7.4.5 Arches and vaults

Buildings with arches and vaults are a characteristic feature of all historic built heritage. Contrary to a widely held belief based solely on the knowledge of the modern technical theory of the beam, which has had negative effects on the conservation of ancient architecture, the behaviour of arches and vaults during earthquakes is much better than is thought, as is shown by analysis of the ensuing damage. It is no coincidence that basements and ground floors are traditionally covered with vaults to increase the box-like behaviour of the building.

Throughout the Mediterranean world, arches have been used as an anti-seismic device to join various buildings in order to dissipate seismic energy over large volumes or to join highly asymmetrical parts of buildings.

In ancient masonry roofs with a flat extrados and a vaulted intrados, resistance is ensured by the whole of the masonry and not just by the arch-shaped intrados, enabling the curve of the pressures to adapt to the various stresses.

With regard to the presence of crack patterns, localised supplementary interventions can be carried out on arches and vaults in much the same way as masonry, with the proviso that some cracks may be physiological, so that it may be better not to repair them; this is particularly true for cracks in keystones and skewbacks as long as the resistance to the thrust is ensured.

The judicious use of chains, of the type already described, can bring about significant improvements; they should normally be placed at the "skewbacks" of arches and vaults.

It is advisable to avoid using plate end anchorage (e.g. using fibre-reinforced polymers), create a counter-vault at the extrados; these techniques make the structural behaviour of the two overlying elements ambiguous and create unevenness between the materials which have extremely different degrees of durability; lastly, they are practically irreversible.

7.4.6 Floors

Floors are normally subject only to vertical loads. When earthquakes occur, they play an important role in linking the masonry walls and as transverse stiffeners, helping to create the "box-like effect" which is essential to the spatial behaviour of the building. For this reason, floors need to be firmly connected to the load-bearing walls and should be adequately rigid in their own plane.

Given this, the following courses of action should be undertaken. Creating stringcourses that lead to continuous stress on the masonry should be avoided. Regular perimeter links should be established and, where necessary, chains and continuous supports in wood or structural steelwork which are fully integrated with the masonry through regular links. If the floor needs to be replaced, the intervention must be largely reversible; wooden floors should normally be preserved but, if they do need replacing, they should be substituted with another wooden floor. The ends of the beams should be firmly embedded in the masonry and rest on stone or brick sockets; where required, they should be reconstructed, and any signs of deterioration along the beams must be eliminated. It is also advisable, where possible, to add an additional bond or tie that prevents major horizontal movement. Lastly, the known techniques for stiffening plank floors should be adopted. It will also be necessary to allow adequate ventilation to prevent the risk of rotting over time; for floors with beams and hollow flooring blocks, it is advisable to carry out the stiffening using a reinforced concrete floor linked to the structural steel beams, prepare chains and protect the structural metal beams from corrosion; concrete and masonry floors should be replaced with new floors, preferably made of wood.

7.4.7 Staircases

Old staircases are generally extremely high-quality parts of a building. They must be preserved, eliminating any signs of dilapidation and restoring damaged parts with masonry techniques and ensuring, in particular, the link with perimeter masonry. In cases of extreme necessity, it may be necessary to use external structural steelwork that reflects the original configuration of the staircase.

7.4.8 Roofs

Wooden roofs with flat or concave roof tiles should be preserved; sometimes it is necessary to eliminate thrusts with suitable tie-beams.

7.4.9 Other interventions

Ruined elements may sometimes have extremely precarious configurations where equilibrium is ensured to the limits of stability, or remains sufficiently stable in ordinary situations but extremely precarious in the event of an earthquake, with masonry elements that are too thin, walls that are too high or not braced and the ruins of vaults which act as cantilevered elements. These problems can generally be resolved with traditional improvement mechanisms, such as tie-beams or steel cables, or with small integrations made using homogeneous materials which are highlighted or, lastly, by restricting access to areas not affected by the risk of collapse. When designing such interventions, the identity of the ruined structure and its material integrity should always be guaranteed.

7.5. The contribution of innovation: Diagnostics and materials

As part of a historical and scientific theory of the conservation of historic built heritage, diagnostics has become an increasingly important tool for our scientific understanding of material history. Geotechnical, structural, physical, chemical and biological diagnostics have recently undergone significant development which will continue in the immediate future, also involving fairly secure dating of inorganic materials. Before planning conservative intervention, besides researching the history of the building concerned, there should ideally be a diagnostic project adapted to the specific case which will lead, partly through accurate mapping, to detailed scientific knowledge of the design of the ruined building, the materials and techniques employed and the various forms of dilapidation. It is also crucial to gather previous monitoring data and analytical studies, which are of great importance for interpreting ongoing phenomena related to instability.

Over the last decade, materials engineering has developed numerous innovative materials such as artificial stone, fibreglass elements, composite materials etc. These materials should be used with great caution in intervention on the built heritage since neither their compatibility with traditional materials, their reversibility nor their durability has been tested. However, these materials are always preferable to traditional forms of consolidation. It should be emphasised that it will be increasingly possible to design materials with specific physical, chemical and mechanical properties so that a further kind of scientific interaction between constructive conservation and engineering can be envisaged.

7.6. Case studies

Unfortunately, when undertaking interventions, the application of the previous recommendations and, more generally, the principles of conservation theory need to take account of a complex set of aspects such as administrative and economic factors, or the interaction with contractors and the various technical departments often linked to traditional forms of consolidation.

All this may inevitably lead to some compromises which scrupulous and well-prepared architects or designers will try to restrict as far as possible.

The following illustrates several examples of conservation and restoration projects which have tried to follow conservation theory closely.

7.6.1 The temples and walls of Paestum

In Italy, the ancient city of Paestum, which was founded by the Greek colonists from Sybaris in the late seventh century BC and is surrounded by large walls, has still only partly been brought to light. Its exceptional nature stems from the three large Doric temples – the Basilica, the Temple of Athena and the Temple of Neptune – which have survived in an exceptionally well-preserved state. The three temples have been subject to major conservative restoration which began in the late 1980s and ended in the early part of the twenty-first century.

The research took several years and used the Temple of Athena as the sample. All possible themes of study were explored, including geotechnical analysis, constructional analysis, the various kinds of vulnerability, the study of lichens and microorganisms and a careful study of the surfaces, which displayed traces of the original colours. This phase ended with a documentary exhibition and a seminar of studies held at the national archaeological museum of Paestum (AA.VV. 1993). Conservative conservation was subsequently carried out over a period of about ten years (Cipriani and Avagliano 2007).

The case history of previous work brought to light the abandonment of the ruins which lasted throughout the eighteenth century and the first precise and effective interventions made by Bonucci in the early nineteenth century using bricks and dowels. This was followed by the interventions carried out by Maiuri during the 1930s, marked by metal cladding and some concrete inserts, which were fortunately limited in extent. In 1962, this was followed by the rash intervention by Fondedile with metal cramps inserted of the lintel in the Temple of Athena and the placement of metal bars in the columns of the pediment in the upper part of the same temple. Shortly afterwards, a column of the pediment was struck by lightning, attracted by the steel, partly destroying the first drum of the column; this was subsequently patched up with the probable use of bolts.

The various interventions described above clearly reveal a distinction between initiatives that respect the constructional approach of the ancient buildings and those which, more or less deliberately, contradict it, imposing, not just on the elements of restoration but also on the archaeological ruins, static and structural behaviours which are completely at odds with the original ones.

The consolidation interventions can be placed in at least four categories, which are homogenous in terms of the use of materials and techniques.

A. The first category concerns interventions where the collapse has been corrected through partial dismantling and reassembly of the stone blocks, with the replacement of the missing or damaged parts using blocks of material that are similar to the original. The reconstruction of a central part of the western trabeation of the Temple of Neptune, probably damaged by lightning, is emblematic of interventions of this kind (Figure 7.21).

Figure 7.21 Paestum, the Temple of Neptune and the central part of the western trabeation – Photograph by the author

In these cases, the executive technique, which carefully mirrors the extraordinary precision of the ancient construction, is extremely interesting. These interventions require the replacement and/or reworking of the original material and the dismantling and reassembly of relatively large portions of the ancient building. Overall, restorations of this type that can be found on the temples are extremely well-preserved.

B. The second category comprises interventions designed to consolidate, fill in gaps and safeguard elements of buildings that are unstable, carried out using bricks and mortar. An interesting example of this kind of work is the nineteenth-century restoration carried out by Bonucci on the Temple of Athena (Figure 7.22). This type of intervention involves the use of a technique and materials which, despite introducing fragmentary elements of discontinuity into the historic building, respect the overall construction techniques. Although they differ from those of the original building, the materials employed for consolidation do present significant analogies, both in terms of structural behaviour and in terms of durability. The additions in brick and mortar to fill gaps are clearly distinguishable from the historic building. The new brick masonry is carefully adjusted to the sections and profiles of the existing gaps, making it possible to reduce as far as possible, and sometimes avoid altogether, cutting or dismantling of the original stone blocks.

The restoration work of this type carried out on the temples of Paestum is well-preserved and is in perfect condition in terms of structural efficiency. At some points it is possible to notice a few small gaps and elements of discontinuity with the use of the same materials and the same building technique.

Figure 7.22 Paestum, restoration carried out by Bonucci on the Temple of Athena – Cipriani, Avagliano: Il restauro dei Templi di Poseidonia

C. A third category of consolidation intervention include all types of work, consisting of metal elements, aimed at inserting hoops and chains and generally trying to secure individual elements of the old building or groups of them. A particularly significant example of this kind of intervention, which is present in all the structures under consideration here, is to be found in the Basilica. The elements used in this case are made with iron sections and plates which have been carefully shaped, soldered and bolted so that they can be adapted and fastened to the profiles of stone blocks that need to be secured (Figure 7.23).

In some ways, restoration work that falls within this category betrays the compositional and structural sense of the ancient building, although it displays striking autonomy, underlining its formal and structural independence. From both a conceptual and an operational perspective, these types of intervention can be considered to be types of temporary support that do not form part of the ancient building. They are completely reversible interventions and, as the history of restoration at Paestum clearly shows, can be easily replaced and maintained.

D. The fourth category concerns recent interventions carried out following the introduction of reinforced concrete. They were carried out by Maiuri on the Temple of Neptune, both on a perimeter column and on the double order of columns of the cella. Further work was conducted on the Temple of Athena by Fondedile, clearly illustrated in the examples of consolidation intervention.

These interventions, which are understandable at a historical level, are the result of a moment of "confusion" between the introduction of a new material and the lack of attention to material history, but are completely unthinkable in terms of respect for conservation

Figure 7.23 Paestum, restoration through shaped, soldered and bolted iron sections and plates, ("Cerchiature di Maiuri") – Cipriani, Avagliano: Il restauro dei Templi di Poseidonia

theory and in terms of a far more mature knowledge of the technological qualities of reinforced concrete.

7.6.1.1 Improvement interventions

The analysis of vulnerability, the safety and stability of the original building, the crumbling of ruined structures and the static calculations required for particularly severe conditions which are not immediately threatening reveal that the current state of static and structural preservation of the three temples is excellent and the overall safety coefficient during seismic activity is still very high (D'Agostino and Frunzio 1994). This result has recently been confirmed by research carried out by the universities of Salerno (Italy) and Kassel (Germany) (Zuchtriegel et al. 2019). The static restoration has rigorously followed the concept of improvement in the sense of the overall complex of interventions linked to the elimination of structural dilapidation without intervening in the material conception of the monument. The intervention then focused on controlled weed-killing and the elimination of animal and plant parasites, and careful cleaning of the stone parts and subsequent plastering. In particular, when cracks were detected in the ashlars, repointing was undertaken using non-cement-based mortars that have great mechanical resistance and durability. The few metallic parts, generally consisting of external supporting elements, were recleaned and, where necessary, underwent structural restoration and appropriate treatment to reduce dilapidation over time. The areas most exposed to weathering were treated with carefully designed waterproofing, which left the ancient stone masonry intact (Figure 7.24). Lastly,

Figure 7.24 Paestum, restoration by system of waterproofing in the Temple of Athena – Photograph by the author

access to the inner part of the temples was restricted to prevent dilapidation caused by prolonged wear and any possible acts of vandalism. Unfortunately, this intervention, which was very respectful of the monuments, has recently been eliminated in favour of a poorly understood concept of enhancement.

Such accurate conservation of the masonry of the temples had never been previously attempted. As already demonstrated, the masonry structure has high safety coefficients and can now confidently face the challenges of the future and be entrusted, in its current integral state, to new generations.

7.6.1.2 THE PROGRAMMED MAINTENANCE PROJECT

However, in order to reduce conservation interventions in the future, it is advisable to plan cycles of programmed maintenance.

An accurate visual analysis from the ground with the aid of suitable binoculars is required at least once a year in order to record the onset of small cracks, which should be carefully mapped. Every three years, a wide-ranging survey should also be carried out of the upper parts with the aid of a forklift to check the state of preservation of the stone masonry and the weatherproofing layers. Work will then be done to eliminate weeds and, where necessary, the weatherproofing layers will be reapplied, stuccoing will be done and, if required, small cracks will be repaired, assessing the state of preservation of the metallic parts and, where necessary, protecting them from oxidisation.

Clearly, if an earthquake occurs, the survey will be carried out immediately afterwards. If the cycles of maintenance described are carried out, it will be possible to put off the need for more extensive conservation for several decades.

Unfortunately, after several years, the drum of the pediment of the upper part of the Temple of Athena, which had been struck by lightning, began to display a series of micro-cracks, due probably to the metal rivets used to patch it up. An accurate diagnostic analysis showed that the phenomenon has continued very slowly and is tending to come to an end. Several alternative solutions have been suggested to contain or eliminate the phenomenon which, unfortunately, have not been implemented (Brigante 2018). The Istituto Centrale per il Restauro e la Conservazione has recently carried out important research to monitor the state of preservation of the temples (De Palma 2018).

Intervention of a similar nature to that undertaken on the temples has been carried out on several stretches of the ancient city walls of Paestum, revealing its complex constructional organisation. Many blocks have been repositioned while those in an uncertain position have been surveyed and classified, preserving them behind the walls. In this case too, it is crucial to conduct programmed maintenance, without which the state of preservation will decline rapidly.

7.6.2 Villa of the Quintilii on via Appia in Rome

Villa of the Quintilii, situated at the seventh milestone along via Appia, occupies an area of 23 hectares; the original owners were the Quintilii brothers, who lived in the late second century AD, but the attribution to the family dates to 1828, when several pieces of lead pipe (*fistulae*) were found with the inscription *Quintilii Condianus et Maximus* (Figure 7.25).

The Quintilii family, who were wealthy landowners, were connoisseurs and writers of works on land surveying and military topics. Thanks to the protection of the emperor

Figure 7.25 Rome, Villa of the Quintilii on the via Appia – Photograph by the author

Marcus Aurelius, they had dazzling political careers: they were governors of Acaia (Greece) and Pannonia (present-day Hungary), where they managed to thwart attempts at invasion by the Germans; they both held consular posts in AD 151 and, once again together, were sent to their deaths by the emperor Commodus in about AD 182.

Between AD 182 and 183, Commodus seized the villa, whose fertile land and the presence of water made it a site that was coveted and frequently visited by the emperors.

The central part of the villa dates to the age of the Quintilii family in the late Hadrianic period, but its transformation into an imperial residence led to extensive enlargements; numerous finds have demonstrated that the villa remained in use until the third century AD.

There is very little evidence for the use of the area over the following centuries. During the Middle Ages, some parts were fortified and, by 997, the area belonged to the Celimontano monastery of di S. Erasmo. The villa appears in the map of the Ager Romanus drawn up by Eufrosino della Volpaia in 1547. During the seventeenth and eighteenth centuries the villa was given the name Roma Vecchia and in 1797 it became the property of the Torlonia family. As a result of the discovery of huge numbers of statues and archaeological finds, it was also known as Statuario.

The first intervention dates back to 1485 when a sarcophagus was found with the body of a girl wearing a gold diadem in her hair. Other finds are mentioned by Winckelman, and some statues that may come from the villa are now on display in the Hermitage in St Petersburg.

During the second half of the eighteenth century, many discoveries were made, and the finds are now in Munich, the Louvre and the Vatican Museums.

In the nineteenth century, excavations were carried out by the Torlonia family, who enriched their own palaces beyond the Museo Torlonia, and the villa was finally attributed to the Quintilii brothers.

A history of the excavations and a sufficiently detailed analysis of the various monuments is contained in a recent volume published following a project to excavate and enhance the site carried out by the Archaeological Superintendency of Rome between 1998 and 2000.

> The results of these excavations were surprising and have shown that further exploration is required to reveal the plan of the complex and explain the function of the rooms in the various sectors. To enable visits to the main parts of the villa, it was decided that the most "monumental" buildings should be investigated while, at the same time, following the priorities of archaeological exploration and restoration, ensuring that these initiatives could be combined with safeguarding the landscape which has remained virtually intact, although it must be adapted to receive visitors.
>
> [Paris 2000]

Now that this project has been rapidly completed, Villa of the Quintilii is one of the most beautiful and intriguing archaeological sites of ancient Rome and is worth a visit in its own right.

The grandeur of the some of the monuments is second only to the Baths of Caracalla. The residential sector with reception rooms, the baths with the splendid Frigidarium and Calidarium, the huge cistern and the large nymphaeum, are famous monuments whose restoration has been significantly influenced by conservation theory and has been particularly respectful of the ancient fabric of the buildings through the use of bricks and non-cement-based mortars. Some of the most interesting interventions are described below.

The south-east side of the large room of the frigidarium has a large rectangular window, about 8 metres long and about 5 metres high, which, in the upper part near the corners, has two incomplete arches and a long, virtually straight intrados (Figure 7.26). Above the window, there are clear signs of the insertion of a barrel vault roof, which has been destroyed. Its ruined remnants are situated in the zone that has yet to be excavated in front of the wall. The existing records showed that the situation had remained unchanged for about a century, and the masonry did not display any significant signs of collapse besides a minor series of micro-cracks (Figure 7.27). Several solutions were proposed, although all of them misrepresented the archaeological interpretation. The current author demonstrated the possibility of leaving the current situation intact since the simulation of the finished elements revealed the presence of extremely modest shear stress that is completely compatible with the nature of the masonry, which was improved by inserting micro-injections between the mortar joints. Long-term monitoring of the wall has confirmed the wisdom of this choice. In the frigidarium, covered with a cross vault whose massive springers can be seen, the north wall has a large arch whose supporting walls have largely collapsed. Limited restoration work has been carried out on the masonry, leaving the form of the ruins and thus the collapsed masonry clearly visible, only ensuring the stability of the ruined parts, respecting the principle of minimum intervention (Figure 7.28). The slow and inexorable evolution of the abandoned ruins led to the formation of a large portal consisting of two massive masonry pillars supporting an arch whose central part had collapsed. Static testing ensured the stability of each part not connected to the keystone; in this case, the archaeologists suggested completing the arch with brick masonry which was entirely compatible, in aesthetic terms, with the ancient fabric (Figure 7.29).

The large room of the Calidarium (Figure 7.30) had fallen into a state of ruin despite preserving a striking masonry structure with a towering height of over 20 metres. On one

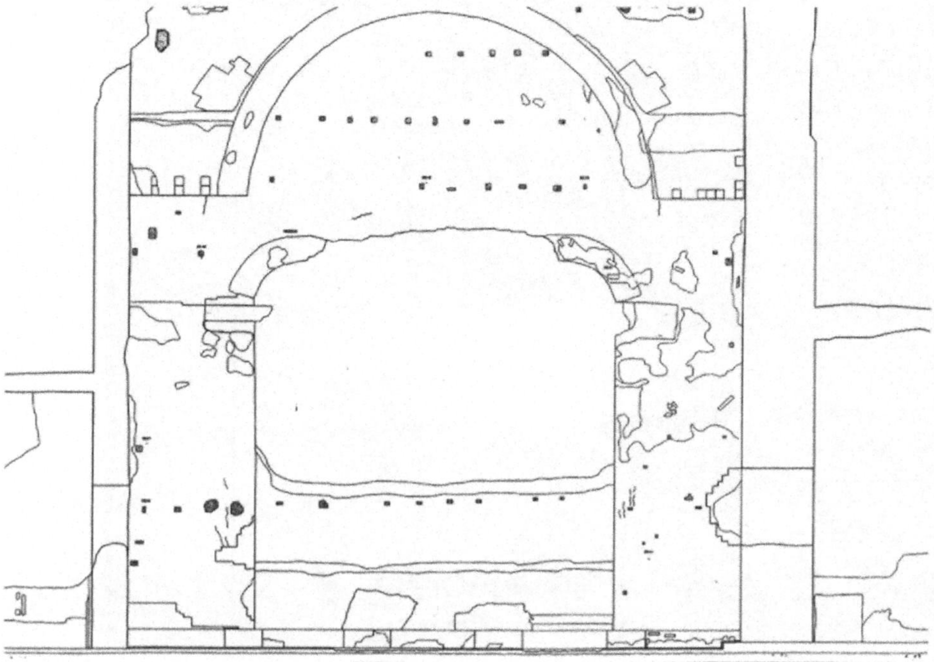

Figure 7.26 Villa of the Quintilii, diagram of the south-east side of the Frigidarium – Drawing by the author

side, a wall has a solid connection with the orthogonal masonry while, on the other side, it is completely free. Its stability is ensured in ordinary day-to-day operations, but it could partly collapse in the event of an earthquake. Instead of tampering with ruins through inappropriate additions, an assessment was made of the possible extent of the collapse, closing off the area in order to exclude access to visitors.

Particular attention was focused on restoring the roofing of the large cistern, carefully adjusting the ancient extrados through minimal restoration of the masonry, additions and stuccoing, while precise restoration of the interior was undertaken using patching where necessary or the simple insertion of brick fragments in all the vaults.

The restoration of the large nymphaeum on the ancient via Appia is worth discussing in more detail (Figure 7.31). The masonry of the nymphaeum is made of *opus vittatum mixtum*, which consists of a course of parallelepiped tuff blocks and a course of bricks (Figure 7.32). Its construction is dated to the second half of the second century AD, possibly during the reign of Commodus (AD 180–192). The same technique is found in several parts of the main building and was already in use in the first half of the second century AD, even though it became more widespread in the later imperial period (Frontoni 2000).

Medieval interventions repaired the damaged ancient masonry, using salvaged material of various kinds (tuff, brick and flint) and added new buildings designed to defend and control the road, using masonry of parallelepiped blocks of peperino, laid in regular courses with dark-coloured pozzolan mortar.

Figure 7.27 Villa of the Quintilii, proposed reinforcement of the façade of the Frigidarium – Photo and drawing by the author

Figure 7.28 Villa of the Quintilii, Frigidarium – Photograph by the author

Figure 7.29 Villa of the Quintilii, detail of arch with brick masonry compatible with the ancient fabric – Drawing by the author

Figure 7.30 Villa of the Quintilii, Calidarium (perspective view) – nineteenth-century print of watercolour by Labruzzi

Figure 7.31 Villa of the Quintilii, the Nymphaeum – Photograph by the author

Figure 7.32 Villa of the Quintilii, a detail of the masonry – Photograph by the author

Between 1850 and 1851, Luigi Canina directed a huge campaign of restoration and reconstruction of via Appia, but the nymphaeum was restored later on between 1909 and 1913 by Antonio Muñoz, who redeveloped and patched up portions of crumbled masonry without repairing the vaults that had broken and collapsed; no excavations were carried out during the intervention. The gaps in the opus vittatum were filled using similar masonry which was laid without any undercut, replacing the yellow tuff with grey peperino.

The ancient and later masonry displayed widespread problems of detachment and collapse; the points of contact between the medieval and the Roman masonry showed significant detachment and apertures due to the lack of toothing and pressure from roots and weeds; localised situations had also deteriorated due to the flow of rainwater.

The collapse of the vaults, which are important structures ensuring horizontal connections, has, in many parts of the building, doubled the number of free-standing walls which now lack the medieval wooden floors and ceilings.

The building had fallen into a state of ruin, which had seriously undermined the constructional logic of the original structure, altering the balance of the masonry masses and the mutual interconnections.

Rather than seeking to reconstruct the many collapsed vaults and arches, the restoration work began by selecting the main weak points. The decision was then taken to restore the structural continuity at the points where such intervention was deemed necessary in order to improve the general stability of the structure. Only one vault was repaired (Figure 7.33) in the room between the hemicycle and the garden; the walls that had become detached from

Figure 7.33 Villa of the Quintilii, an example of a repaired vault – Photograph by the author

adjacent buildings were all reconnected, the detached parts were toothed and all the tops of the walls were repaired (Figure 7.34).

The edges of the collapsed vaults, which had been seriously damaged by rainwater and the impoverishment of the mortars, were meticulously repaired, and inner cores were injected to improve their consistency (Alberti *et al.* 2004).

Further intervention in the future will involve the reconstruction of several wooden floors belonging to the fortalice which, with a waterproofed extrados, would protect the cisterns below, currently exposed to the elements.

The left corner of the vault in the large exedra displayed a significant detachment of the arch built of bipedal bricks; the cracks were injected with liquid mortar, and the toothing of the arch with the concrete core of the vault was repaired with restored bipedal bricks, creating, at the back, a small buttress to improve the bracing of the reconstructed part of the vault.

The restoration work ensured the careful adjustment of all the loosened masonry, fracture lines and detachments of parts of the masonry, which had drastically reduced the safety levels of the walls; the building regained its original equilibrium, and the masonry was not transformed through the addition of elements or materials that would have been incompatible with the existing ones. The traditional materials were used with great expertise by the artisans and stone masons. The existing architectural structure, each part of which was carefully studied, guided the decisions regarding intervention, bringing about widespread improvements in the structures of the nymphaeum of the Villa of the Quintilii (D'Agostino and Bellomo 1999).

Figure 7.34 Villa of the Quintilii, an example of repaired walls – Photograph by the author

Figure 7.35 Hadrianic façade of the Tiberian domus, general view from the Roman Forum – Laser scan survey by P. Gasparri CPT Studio, 2018

7.6.3 The case of the Palatine Hill and the Domus Tiberiana in Rome

7.6.3.1 The Palatine Hill

The Palatine Hill is a unique site in terms of its archaeological importance and the complexity of the conservation problems, especially with regard to the approach and the aims of the interventions.

The area has been continuously inhabited since the Iron Age (Villanovan and Etruscan culture) and was the site of monumental civilian and religious works of architecture for the eight centuries of Roman rule, was vandalised and looted for over ten centuries and was finally partially restructured with new buildings and occupied by gardens and orchards in the seventeenth century until the first attempts at conservation work in the nineteenth century. The end result is a continuous, disorderly stratification of buildings that have been partly damaged, demolished and altered, interacting with each other and with the land on which they are built. Their identification and philological interpretation is fairly simple through the recognition of building typologies and the materials employed, although their conservation and consolidation pose complex interdisciplinary problems and involve difficult, frequently subjective decisions which are, nevertheless, debatable and the subject of constant criticism.

The geological and morphological features of the hill have undoubtedly increased its appeal and have led to the current situation. The hill was formed by the stratification of pyroclastic deposits from the Monti Sabatini and Alban Hills volcanoes: tuff, pozzolan and volcanic sands. Tuff is a soft rock which is easy to quarry and carve to make building material. For many centuries, since the prehistoric era, squared-off blocks of tuff were used in buildings in Lazio and Etruria, while large layers of tuff were used to build settlements or funerary structures, such as the Etruscan cemeteries, in caves and tunnels. Pozzolans and volcanic sands have special mechanical properties which enable natural faces, even very steep ones, to remain stable for considerable heights. The combination of these factors has led to the truncated pyramidal shape of the Palatine Hill, with its steep slopes, facilitating the excavation of channels, caves and tunnels which are independently stable.

Between 2009 and 2013, the geological and geotechnical records, instrumental testing and surveys carried out in the area between the Roman Forum and the Palatine Hill were completely reviewed. This extremely important piece of documentary heritage, kept in the Fondo "Vittorio Ascoli Marchetti", is now kept in the archive of the Archaeological Park of the Colosseum.[1]

The *Domus Tiberiana*, which occupies an area of 32,000 square metres, over a quarter of the surface area of the Palatine Hill, has been recognised by modern historiography as the first great palace of the Caesars: a sumptuous residential complex, consisting of porticoed gardens, baths, reception rooms and large ancillary rooms. The complex stands between the northern slopes overlooking the Forum along *Via Nova* and the southern slopes overlooking the sacred area of the temples of Cybele, Victoria and Apollo, and between the western slopes towards the Velabro valley and the Clivus Palatinus, the ramp running from the Forum to the Palatine Hill.

The ruins of the palace attributed to Tiberius (AD 14–37) actually date to the reign of Nero (AD 54–68), and are partly built on the ashes of ancient aristocratic villas of the Republican period, destroyed during the fire of July AD 64. The ruins of ancient workshops, probably of Republican date, are still visible at the base of the north-eastern face of the Palatine Hill, on the edge of the Forum, where an important drainage pipe from the Etruscan period was also found.

During the reign of Domitian (81–96), the palace was joined to the valley of the Roman Forum through the construction of a massive ramp, while during the reign of Hadrian (114–138), a huge avant-corps was built into the façade overlooking the Forum, with a covered ramp known as the *Clivus Victoriae*, which was linked to *Via Nova* (Figure 7.37).

The support structures of the hill face, the vaults and arches that cross *Via Nova*, date to a later phase, presumably the reign of Septimus Severus (193–211).

From the early fifth century AD, many of the buildings on the Palatine were abandoned and partly destroyed by the terrible earthquake of AD 443. The entire complex of the Domus Tiberiana continued to be inhabited and occasionally restructured. It was used until the seventh century when the Domus was chosen as the seat of the bishop of Rome, Pope John VII (AD 650–707).[2]

Over the following centuries and for the whole of the Middle Ages, the Palatine was gradually abandoned and the palace was continuously stripped, both of its most precious components and materials such as columns, capitals and floors made of stone, and of poorer materials, such as tuff and pozzolan. When Cardinal Alessandro Farnese commissioned Vignola to create the Farnese Gardens in 1564, the remaining structures were safeguarded and only partly adapted to new functions. Over the subsequent centuries, the hill was practically abandoned until 1870, when the first archaeological excavations were undertaken by the Italian state.

Although the major excavations carried out by the French administration and continued after Italian unification brought to light striking archaeological remains, they also profoundly altered the morphological and architectural state of the imposing Hadrianic sub-structures

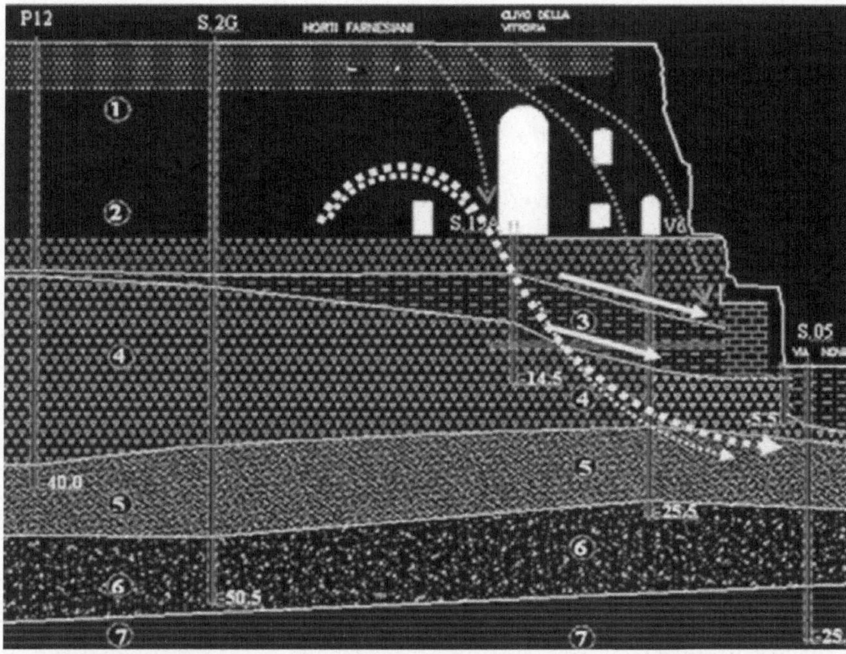

Figure 7.36 Geological and architectural section of the sub-structures of the Tiberian domus. The direction of the downward slide of the structures is shown with hatching – Drawing by V. Ascoli Marchetti and G. Cangi.

Figure 7.37 Via Nova – Photograph by L. Filetici

towards the Roman Forum. Demolition was not followed up by the necessary consolidation work, which was only partly undertaken from 1950 onwards.

A new systematic general project designed to safeguard the façades of the Domus Tiberiana was begun after an initial phase of research and testing in 2006, followed by major restoration work which started in 2009 and has continued up to the present. The work has solved much of the structural fragility of the Hadrianic part of the complex straddling the *Clivus Victoriae*. The diagnostic phase, the monitoring and multi-disciplinary interventions continue within the context of a general overview of the imperial palace and its relationship with the surrounding archaeological areas, and major economic investment to create the Archaeological Park of the Colosseum.

The interventions carried out so far are too extensive and numerous to be discussed in detail here. However, they all share a common methodological approach which has been pursued until now, based on the interpretation of the results of both previous and recent research, deemed to be crucial for an understanding of the nature of the movements recorded along the slopes of the hill towards the Roman Forum and the Velabro valley.[3]

Geotechnical investigations and structural monitoring have shown that the recorded movements did not stem (as was feared in the past) from the instability of the clayey layer beneath the Hadrianic structures, but rather from general subsidence of the area due to consolidation, caused by the varying conditions of underground water courses. Work was therefore carried out to consolidate the masonry, vaults, arches and lintels to repair the cracks and restore the structural failings caused by collapses and plundering.

Besides the major earthquakes of the past, the causes should also be attributed to the looting of ancient materials over the centuries and the building transformations which have modified the architectural fabric of the original complex over time.

Following structural, geotechnical, hydrogeological and physical tests, areas were selected where it was necessary to intervene in order to repair the extensive damage to the masonry, which had led to instability and presented dangers.

The restoration work can be defined as an example of *structural anastylosis* by which, following case-by-case assessment and adopting an approach based on minimal intervention, the reciprocal actions between vaults, arches, lintels and masonry have been re-established in cases where various causes had impeded correct static functioning.

The materials adopted for the interventions were the same as those used in the construction of the original complex. The natural materials, mainly tuff together with bricks, lime mortar and pozzolan, are still available and were used by specialist staff following traditional techniques. When it was deemed necessary to improve safety standards and comply with the anti-seismic regulations, or to improve the stability of the slopes of the hill which had been built on over the centuries, tie rods made of stainless steel or fibreglass were inserted, their size depending on the needs that emerged from structural tests (Figure 7.38).

In the light of the above, what follows is a description of the interventions on the north-eastern slope overlooking the Roman Forum and the north-western slope towards the Velabro valley in order to illustrate the key concepts and the methodological approach of the project in accordance with the aims of this volume.

7.6.3.2 North-eastern slope

Via Nova runs along the foot of the slope. From the first centuries of city's existence throughout the imperial period, this road provided access from the east to the civic and religious centre of Rome. An important drainage pipe (*cloaca*) dating to the Etruscan period, covered

Figure 7.38 Seismic improvement work – Photograph by L. Filetici

by large parallelepiped blocks of tuff, was found between *Via Nova* and the area generally referred to as the Roman Forum. On the hillside, several *tabernae* dating to the Republican period overlook Via Nova. Towering over the *tabernae* are masonry buttresses which support the covered ramp known as *Clivus Victoriae* and the part of the palace dating to the Hadrianic period. Further up the hillside are the striking ruins of the rooms of the imperial palace. The fact that these buildings were constructed over such a long period has clearly affected the choice of the forms of intervention for the consolidation of unstable structures, repairs of the cracks and the reconstruction of missing parts.

The instability and collapses of some of the structures originated partly from the previously mentioned earthquake of AD 443, partly from continued looting during the Middle Ages and lastly from the early archaeological excavations of the late nineteenth century. In order to bring the area of the Forum to light, a layer of about 7 metres of deposit, earth and collapsed material, which had contributed to the equilibrium of the masonry buttresses built to support the remaining parts of the palace, was removed during these excavations. A further cause of instability has been identified as the lowering of the soil due to the consolidation of the clayey layer below the tuff layer caused by variations in the underground water levels. Over the last 50 years, the bed of the river Tiber has deepened due to the changing relationship between erosion and deposits, caused by the numerous hydroelectric power stations built on the river to the north of Rome, lowering the water levels in the areas adjoining the course of the river.

A clear sign of this instability is the longitudinal crack passing through the arches and roof vaults of the *Clivus Victoriae*. Interventions designed to consolidate the structures on the slope began in the upper part with the repair of the cracks and the insertion of metal chains between the pillars supporting the vaults. The work was done in various stages, with gaps of several years, simultaneously monitoring the movement of the structures.

To re-establish the structural continuity of the pillars and vaults, the parts that were missing due to collapses or looting of travertine blocks were replaced by brick masonry bound together with lime-based mortar and pozzolan, extremely similar in composition and size to the original masonry. The reconstructed parts were laid so they were recessed in order to ensure they could be distinguished from the original masonry. The reconstructed masonry ensures that the static and anti-seismic behaviour of the overall structure is consistent with the original design without creating imbalances in the distribution of internal stresses. In general, the interventions were restricted to what was considered necessary and sufficient to ensure the conservation of the archaeological monuments and to achieve the legally required levels of safety to enable visits and access.

At the same time as the structural intervention, work was carried out in the rooms situated on the outer side of the covered ramp to remove the collapsed masonry and dumped material, which were undermining the equilibrium of the buttresses and were leading to the movements that had been noted during the monitoring process.

The removal of this material, which was necessary for reasons of stability, brought to light important archaeological finds which had lain buried for centuries beneath deposits of collapsed material.

As the collapsed material deposited at the base of the *tabernae* was removed, a layer of ash several centimetres thick emerged. The ash layer was contiguous in all the rooms and probably dates to the large fire that took place during the reign of Nero in AD 64.

At the base of the Hadrianic façade along Via Nova, work was carried out to reconstruct the missing parts of the vaults of the *tabernae*, of which only the perpendicular walls at the front of the hill were preserved. The intervention (suggested in the past by the late lamented Antonino Giuffrè) made it possible to restore the structural system of the walls according to the Cartesian grid, recreating the original three-dimensional structure designed to contain the hill face which had been deformed and rotated.

The reconstruction of the vaults respected the ancient construction techniques. The structural gaps were paired and integrated with recesses of the planes, pointing of the joints and the choice of bricks which were identical to the original ones in terms of dimensions, construction, type of clay and colour, fully respecting the architectural and structural characteristics of the remaining part of the archaeological complex.

7.6.3.3 North-western slope

On the slope overlooking the Velabro valley, the structural collapses were significantly more serious, and the equilibrium of the face was so unstable that during the 1990s a temporary support structure made of steel tubes was built, which hindered visibility. This side of the Palatine also contains the ruins of structures from different periods, beginning from the first centuries of the Etruscan settlement until the work carried out by the Farnese family in the sixteenth century. The bottom of the slope has important remains of the *Horrea Agrippiana*, the warehouses for storing the grain that came from the nearby river port. They had been brought to light in the twentieth century with the demolition of the medieval residential settlements that were widespread throughout the area between the Palatine and the Campidoglio, of which only the sixth-century church of S. Teodoro remains. At the foot of the tuff face, there are imposing buttresses that reinforce the overlying support structures. A huge wall dating to the Farnese period, which has only partly collapsed, was built against a long stretch of the imperial wall.

In the upper levels, the tuff face is disjointed and subdivided into independent blocks, made unstable by irregular discontinuities created over time by collapses of the slope, full of earth from the gardens above.

On this face too, the consolidation interventions began at the top, removing the earth covering the disjointed tuff blocks and part of the venerable cypresses marking the edge of the Farnese Gardens, the roots of which had delved deep into the cracks. The unstable tuff boulders were gradually reconnected to the stable rock mass behind using a series of ties. To reduce the forces applied to values that were compatible with the low resistance of tuff, fibreglass tie-beams with a small diameter of 40 mm and a length of about 6 metres were fastened only for a brief stretch on the edge. Unlike steel tie-beams, fibreglass does not require protective cementing. Since the extent of the cracking of the rock mass behind the detached boulders was unknown, as was the presence of tunnels and caves, it was necessary to avoid cementing up the holes with the possible risk of altering the conditions of the rock mass, or even causing irreparable damage to the areas of archaeological interest by injecting mortar in uncontrolled quantities.

By carrying out the intervention gradually from the top, it proved possible to dismantle the temporary metallic structure from the whole of the hill face and to remove the earth and stone fragments covering the hill face. The deposits from various collapses were therefore removed from various rooms dug into the tuff. The partially collapsed vaults of the rooms were reinforced where necessary to ensure their safety according to the criteria of minimal intervention. The removal of the collapsed deposits from the hill face brought to light residential rooms from the Republican period decorated with frescoes of exceptional

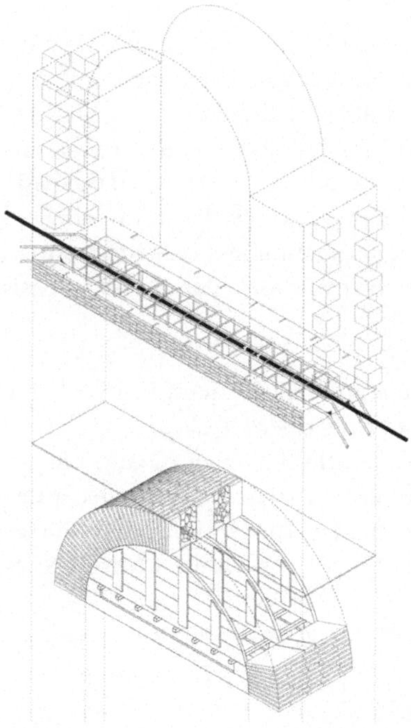

Figure 7.39 View of Domus Tiberiana after restoration – L. Filetici: Drawing

importance that have partly been restored and whose existence was completely unknown. Restoration work on the frescoes is currently being planned (Figure 7.39).

The Farnese support wall and the Roman wall with its buttresses were consolidated through the reconstruction of the missing parts, the excavation of the soil that was pressing against the perimeter wall of the rampart and the creation of a roof to cover the excavation which formed a connection to brace the Farnese wall, which would have otherwise been unstable.

7.7 Conclusions

The need to preserve and consolidate the archaeological remains and historic buildings on the Palatine Hill and make them safe to visit involves issues related to conservation theory, which are more urgent than issues of stability and technology. The buildings date to extremely different periods and are interconnected and interacting. They need to be made safe to visit without altering their form and material construction, partly in order to reveal their original function and the link with the culture that produced them. This makes every conservation intervention an intrinsically difficult problem. The minimal intervention approach that has been followed in recent years for the Palatine Hill is the result of decades of work by numerous scholars working in the fields of archaeology, history, statics, material properties, environmental physics, biology and botany. Together with architects, archaeologists and restorers from the former *Soprintendenza Archeologica di Roma*, now the *Parco Archeologico del Colosseo*, they have worked with great dedication and passion in study groups and committees of experts. The bibliography provides ample proof of their efforts.

7.7.1 The case of Pompeii

Following the devastating earthquake that struck Campania so severely in 1980, initial emergency work was carried out at Pompeii by the military engineering corps, which rapidly shored up the damaged structures with wooden beams. In 1984, the Italian Cultural Heritage Ministry signed an agreement with Società Infrasud to undertake an initial project and to set up a surveillance committee (*Commissione di Alta Sorveglianza*). The committee all agreed on the basic principles of conservation theory. A project was therefore prepared which, based on a heated cultural debate, selected Insula 9 of Regio II as the site for experimental research and testing.

The conservative restoration project of Insula 9 of Regio II was formulated on the basis of the following key criteria:

1. Complete adherence to a single interdisciplinary concept of the project, in which the archaeological part is treated as the guiding element of the project itself;
2. Complete adherence to "Raccomandazioni per gli interventi sul patrimonio monumentale a tipologia specialistica in zone sismiche" (Recommendations for specialist intervention on monumental heritage in seismic zones) prepared by the National Committee for the protection of cultural heritage from seismic activity of the Italian Ministry for Cultural and Environmental Heritage.

The direct consequences of the adopted criteria are:

a. The decision to work towards ensuring "improvements" in the resistance of structures to seismic activity;

b. The decision to use, as far as possible, traditional technologies that ensure the minimum extent of modification to buildings and the maximum amount of reversibility for the proposed interventions.

The adopted criteria and solutions have led to a design process which, by fully respecting the works that survive from the past, tends to avoid distortion in the name of presumed safety both with regard to seismic vulnerability and with regard to conservation, which, ideally, is indefinite. On the contrary, the design process that was followed is based on the knowledge that conservation cannot end in large-scale extraordinary, episodic interventions but should take the form of continuous maintenance that follows the everyday life of cultural heritage as it evolves.

The interventions on various structural elements of the complex of four different houses in Via di Nocera, Insula 9, Regio II are described below (Figure 7.40).

7.7.1.1 Roofing

Roofing follows the styles and typologies established by archaeologists. They can be divided into two types: roofs for which there are clear traces of the ancient roof frame, and roofs for which the general style is clear but the details of the roof frame are lacking. The former are made of natural wood with circular trunks (Figure 7.41), while the latter are made with laminated wood to underline the diversity.

Figure 7.40 Pompeii, the four houses in Via di Nocera, Insula 9, Regio II – Dell'Orto Franchi: Restaurare Pompeii

Figure 7.41 Pompeii, roofing using natural wood with circular trunks – Photograph by the author

Many ancient tiles have been recovered and restored and waterproofing treatment has been crucial. The new tiles are extremely similar in terms of shape and size, but differentiation has nevertheless been ensured (Figure 7.42).

When the load-bearing structure of the roof reproduces the ancient one, it consists of struts with a centre-to-centre distance that is less than the width of the tiles. The struts are securely fastened to the crowning masonry and are treated to ensure long-term resistance to weathering.

Design calculations for large roofs (house 1 and house 4) were done by considering, as a static element load pattern, a flat grid that consists of two frames of principal and secondary beams loaded with their own weight and permanent and accidental loads.

7.7.1.2 Elevations

The height of the masonry varies between 3.5 and 6 metres with a thickness of about 0.4 metres. Their layout defines the geometric structure of the rooms, and they are therefore completely isolated within the boundary wall of the gardens, or more frequently braced within inhabited zones. These walls are often linked to each other but are sometimes set flush with each other. Even though the mortar is in poor condition, often even powdery in the surface zones, the existing masonry does not display alarming crack patterns, except for a few exceptional cases caused by collapses, structural failure of the lintel etc. It was therefore decided that the overall stability of the masonry should be improved in the following ways:

Figure 7.42 Pompeii, the new tiles after restoration – Dell'Orto Franchi: Restaurare Pompeii

1. Complete respect for the existing masonry, and scarification of the joints between the stones using high-pressure water jetting;
2. Removal of cement-based mortar, use of hydraulic mortar with a few additives to ensure in-depth repairs of all the joints that bind the masonry, and for the setting of the flush surfaces, with reinforcement of the joints, where necessary, with wedges of ancient stone or extremely similar material (Figure 7.43);
3. Reconstruction of the sacrificial layer of masonry, which is firmly toothed with the existing masonry in order to restore the masonry to the original height of the roofing. The sacrificial layer will be toothed in all the horizontal intersections in order to create a continuous link as far as possible;
4. Creation of a bond beam made of reinforced concrete wherever possible which constitutes a further connection at the top of the masonry. The bond beam will be placed in the centre of the sacrificial layer of masonry and will therefore not be visible (Figure 7.44).

This approach is designed to leave the ancient masonry intact while reinforcing the joints with mortar that is new but compatible with the original mortar, and to increase the seismic resistance of the masonry through reinforcement and joints at the top due to a sacrificial layer and an overlying bond beam. All these elements are removable. The tests of the analysed masonry, which was subjected to its own weight and accidental excess loads, led to stresses in the materials that can be considered completely safe.

Figure 7.43 Pompeii, reinforcement of the joints with wedges of ancient stone – Photograph by the author

Figure 7.44 Pompeii, the creation of the bond beam in the centre of the sacrificial layer of masonry – Photograph by the author

It has been explicitly underlined that an earthquake, summarised as a static action, is effectively similar to that of wind; moreover, its effects last for a very short time. However, during the test of the boundary wall of the gardens, the maximum stress is 1 daN/cm^2 under normal conditions, and does not exceed 4 daN/cm^2 as a result of an earthquake or wind. For the load-bearing masonry of the houses, the maximum stresses are always lower than the maximum value indicated.

7.7.1.3 Foundations

The situation of places shows that the sub-structure of the masonry is sufficiently stable since it has been compacted over the centuries by excavation material. The base of the masonry is in good condition and the sub-structure is variable according to local conditions. The continuity of the supporting surface creates fairly minor stresses on the terrain, as can be seen from the tests carried out at the base of the masonry.

7.7.1.4 Finishing touches

Some finishing touches improve overall maintenance and contribute to structural conservation. Studies have been done to deal with rainwater which is channelled into collection points outside the insula, or in dry wells within the insula. Samples of *opus signinum* ("cocciopesto", or crushed tiles and mortar) have also been prepared and all the necessary additions have been carried out.

7.7.1.5 Conclusions

The example discussed above is a revealing example of an interdisciplinary approach. By reflecting on the vulnerability and conservation of archaeological sites, the essential role of engineering has been underlined; by reflecting on ancient building techniques, the role of an interdisciplinary approach, in which history becomes the catalyst, has been redefined. This has led to the realisation that some methodological and regulatory approaches stemming from the culture of modern building techniques are at odds with the history and culture of ancient architecture (Franchi Dell'Orto 1990).

7.8 The Citroniera (Orangery) and the Juvarra Stables of the Venaria Reale (Turin – Italy)

7.8.1 Introduction

From 1705 to 1735, Filippo Juvarra's career as a designer and architect took him to the dizzying heights of European art and culture. He designed and built the complex of the *Citroniera* (Orangery) and the Stables between 1720 and 1729 as part of the grandiose design for the Savoy palace of the Venaria Reale (Figures 7.45–7.47). The work reflects the mature phase of Juvarra's career following the completion of the Basilica of Superga and the façade of Palazzo Madama.

His approach to architecture has recently been summarised as follows:

> Juvarra produces his designs by setting out the structural and spatial layout right from the outset with rapid, effective sketches [...]. Spatial invention is never separate from

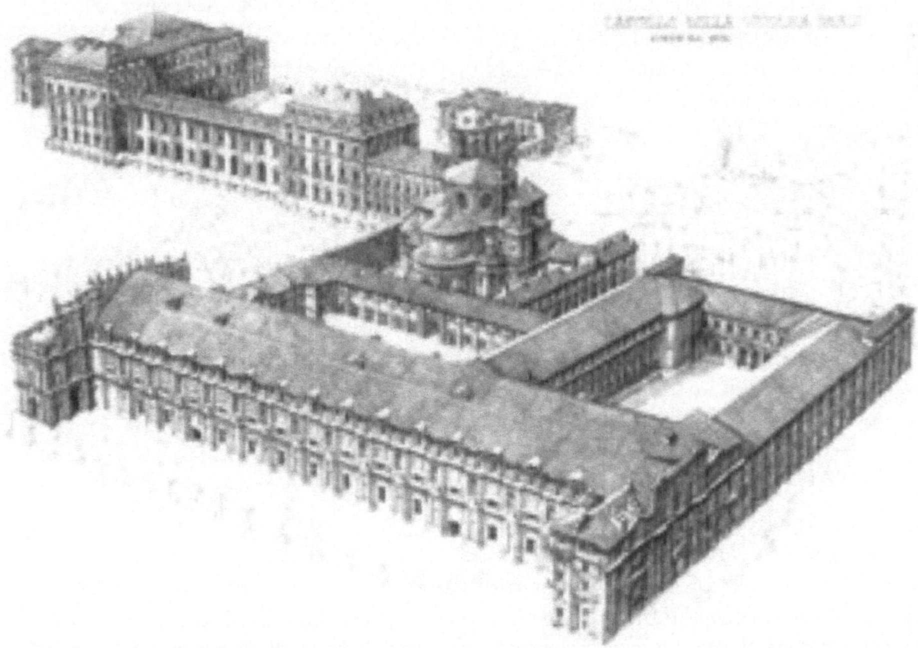

Figure 7.45 Turin, axonometric projection of Venaria Reale - Wikipedia

Figure 7.46 Turin, main entrance of Venaria Reale – Wikipedia

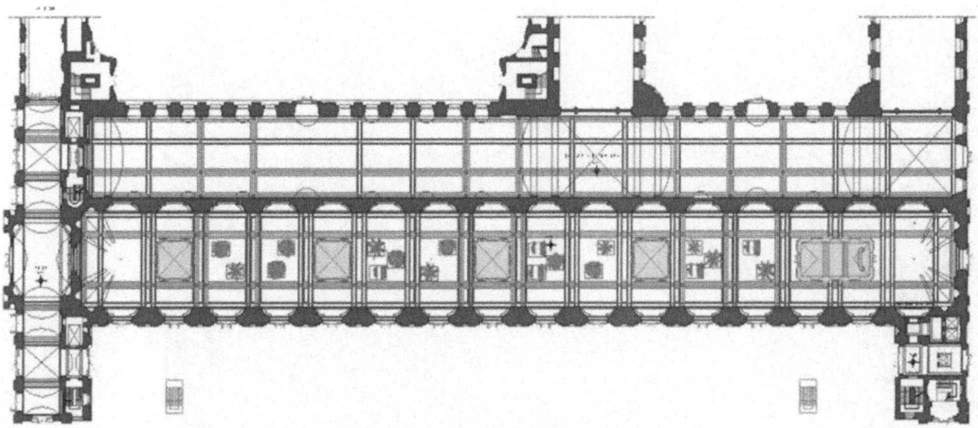

Figure 7.47 Plan of Citroniera (Orangery) and the Juvarra Stables – Drawing by the author

> technical solutions. Rather than a manipulatory set-designer of space, he was a builder who was a "lover of classical compactness" as he liked to define himself; Juvarra regarded technical mastery as essential.

The large complex can be considered as a long rectangle measuring about 150 metres long and about 35 metres wide, divided into two buildings flanked by a spine wall. The two rooms have different ridge heights: the Citroniera is 16.00 metres high while the Stables are 14.80 metres high. In 1814, the Citroniera was turned into stables, and in 1832 part of the palace was transformed into barracks (Figure 7.48).

Its use by the military led to alterations and modifications such as the infill walls built to seal the large windows of the Citroniera and the creation of balconies on the southern façade of the surface area of the dormer windows.

7.8.1.1 The architectural complex: Understanding the building and its history

Juvarra's design was influenced by architectural requirements. The outer wall of the Stables was continuous. The other walls act as mixtilinear pillars connected to each other more or less continuously. At the maximum overall height of about 18 metres, the wall panels have the following thicknesses: outer wall stables, 195 cm.; central pillared wall, 250 cm.; external pillared wall, 300 cm. The hall facing south-west is a large, extremely ambitious work of architecture, also designed in the form of a gallery, although unfortunately it was not completed and was ruined by subsequent interventions. It represents the entrance to the two buildings of the Citroniera and the Stables and has four floors. The hall contains five spans covered by various types of vaults: barrel vaults, ribbed vaults and cross vaults. The two large galleries on the ground floor are covered by a series of brick vaults whose spandrels, corresponding to each pillar-half pillar, become a continuous partition reinforced by the upper buttress and by a complex system of anchor bolts fastened to the vertical masonry (Figure 7.49). The upper floors, both in the Citroniera and the Stables, are of modest height and were used as ancillary rooms. They are covered by a complex system

Figure 7.48 The Juvarra stables in a painting by Carlo Bossoli – Wikipedia

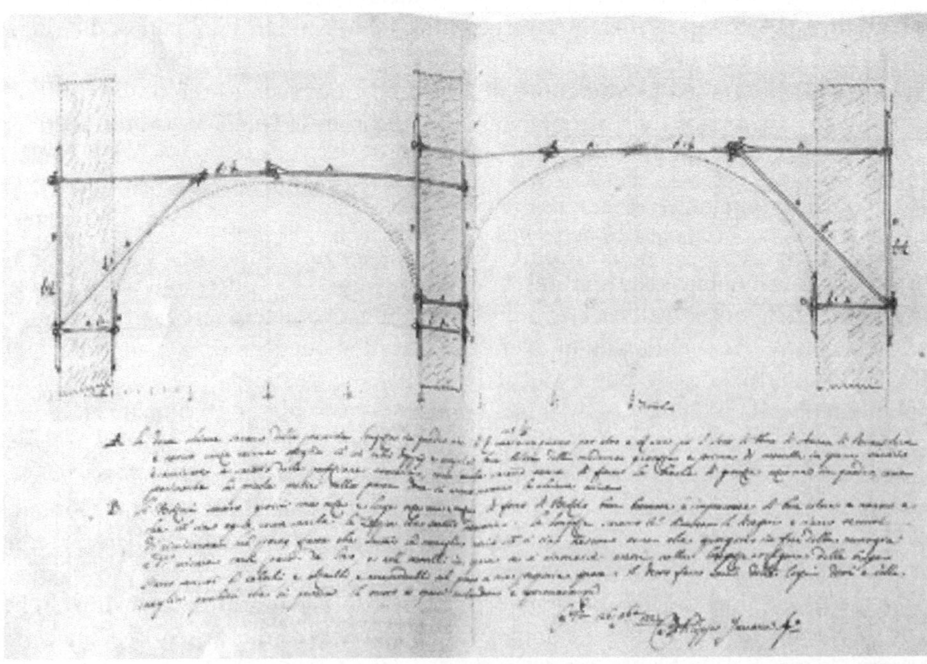

Figure 7.49 The system of anchor bolts fastened to the vertical masonry – Photograph by the author

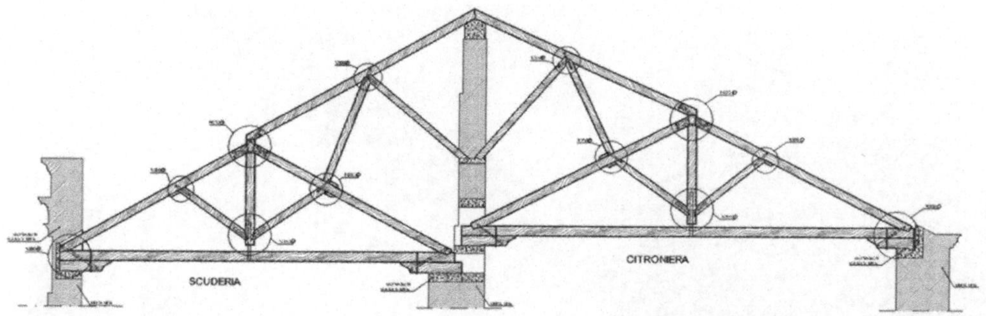

Figure 7.50 System of trusses forming a monumental roof – Drawing by the author

of trusses which form a monumental roof (Figure 7.50). The load-bearing walls continue beneath the ground floor with suitable increased thicknesses, creating an effective system of direct foundations.

The architectural complex has undergone several restructuring interventions caused by the need to convert it to military stables. Some of the most significant interventions include the partial infill of the southern façade of the Citroniera and the construction of a cantilevered balcony with a base of reinforced concrete, which has ruined the architectural elements of the façade. Lastly, the monument was completely abandoned for a long time, although work was carried out in the 1980s to restore the roof, providing protection at least from the rain. Although it had fallen into a state of widespread disrepair, with several serious localised cases of dilapidation, the monument has largely preserved its original architectural structure (Figures 7.51–7.52).

7.8.1.2 Diagnostics and the preliminary analysis of the design

Scientific understanding of the building was achieved by means of the following:

1. Assessment of the geological and geotechnical characteristics of the lithotypes of the plot of land and neighbouring areas;
2. Assessment of the physical and mechanical characteristics of the masonry;
3. Assessment of the state of preservation and the mechanical characteristics of the wooden structures of the roof.

As regards the first point, the following measures were undertaken:

- Coring to a depth of 20 m. with the execution of stratigraphic profiles and sampling for ordinary geotechnical analyses;
- Dynamic penetration tests;
- Excavation pits down to the existing foundation level and archaeological-style tests pits;
- Plate load test on the bottom of the pits (Figure 7.53).

Figure 7.51 Façade of Citroniera before the restoration – Photograph by the author

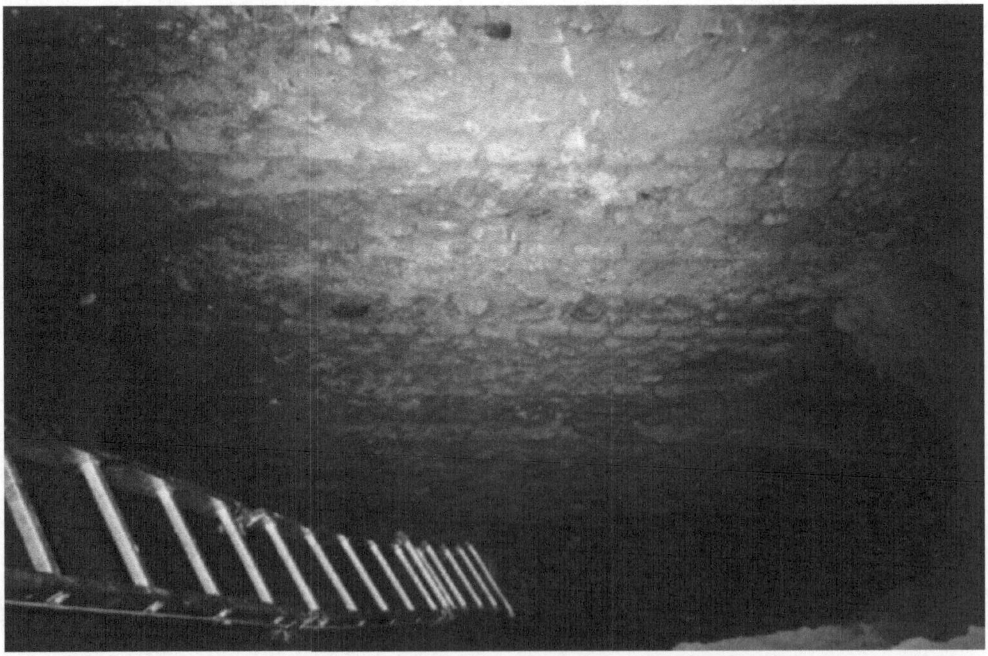

Figure 7.52 Texture of the foundation masonry – Photograph by the author

Figure 7.53 Plate load test at the bottom of the pits – Photograph by the author

As regards the second point, the following measures were undertaken:

- Single and double flat jacks (Figure 7.54);
- Sonic integrity tests;
- Endoscopy:
- Direct tests on the masonry.

As regards the third point, the following measures were undertaken:

- Geometric and thematic survey of the trusses;
- Ultrasound testing;
- Resistance testing (Figure 7.55).

A wide-ranging structural diagnostic campaign was designed and undertaken, making it possible to define the mechanical characteristics of several constructional elements and their state of dilapidation.

The region of Piedmont had carried out a preliminary structural project of the Citroniera-Stables complex in order to assess the feasibility of the intervention that was put out to tender. Before proceeding with the structural improvement project, an analysis of the structural conditions of the monument was undertaken, with a focus on two significant themes:

a) Analysis of the original system of vaults and arches;
b) Analysis of the existing system of foundations.

Figure 7.54 Single and double flat jacks – Photograph by the author

Figure 7.55 Electric testing for resistance measurement – Photograph by the author

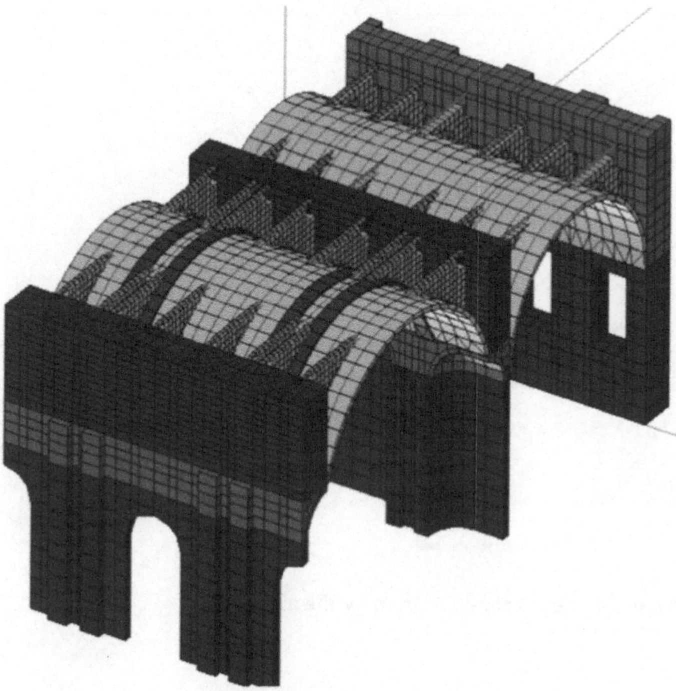

Figure 7.56 Structural model – Drawing by the author

A three-dimensional structural analysis of the calculation model was carried out which consisted of a significant part of the transverse section of the Citroniera-Stables. The aim was to analyse the current operating conditions and to test functionality after the improvement interventions proposed in the project conditions in relation to the future running of the complex. The analysis of the typical transverse section, consisting of barrel vaults measured directly and with known mechanical characteristics (endoscopic, flat-jack and ultrasound tests), are the crucial premise for ensuring the adequacy and congruence of the calculation model with the actual constructional model (Figure 7.56).

- In general, the masonry has an appreciable field of linearity (or pseudo-linearity). Overall, the complex displays linear elasticity. Deviations from the linear field are localised (Figure 7.57).

The results of the static analysis are shown below.

Table 7.1 Isosstresses on grade planes

Wall of Stables	3.26 daN cm^2
Spine wall	7.64 daN cm^2
Wall of the Citroniera	4.69 daN cm^2

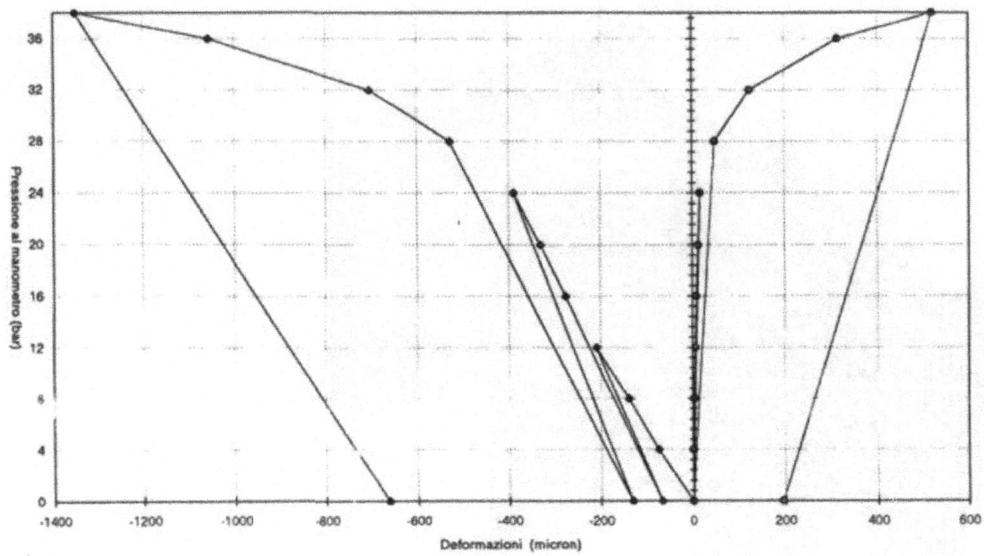

Figure 7.57 Graph of deviations from the linear field – Diagram by the author

Table 7.2 Isosstresses on the springers

Wall of Stables	2.59 daN cm^2
Spine wall	7.39 daN cm^2
Wall of the Citroniera	4.99 daN cm^2
Stress of the chain	1896 daN
Average stress of the vaults	3.21 daN cm^2

The results of the analysis, compared to the results of the diagnostic investigation, clearly indicate that the architectural complex has been preserved in good condition in terms of its stability (Figure 7.58).

7.8.1.3 The structural improvement project

The structural analysis carried out on behalf of the regional government of Piedmont (*Regione Piemonte*) envisaged a possible intervention on the foundations with micropiles, while the diagnostic analysis and the results of the new project have made it possible to avoid altering the original system of the foundations. This is a significant "improvement" since, despite the seismic classification and the new foundations, the adequacy of the foundations built by Juvarra has been demonstrated.

Given the huge spaces of the monument and its environmental setting, an imposing system of underground utilities, built both inside and outside the monument in reinforced concrete, had to be created along the entire surface area of the Citroniera and the Stables in order to house the imposing system of plants and machinery. All the structures are independent of the monumental complex. The only source of interference is the modest number

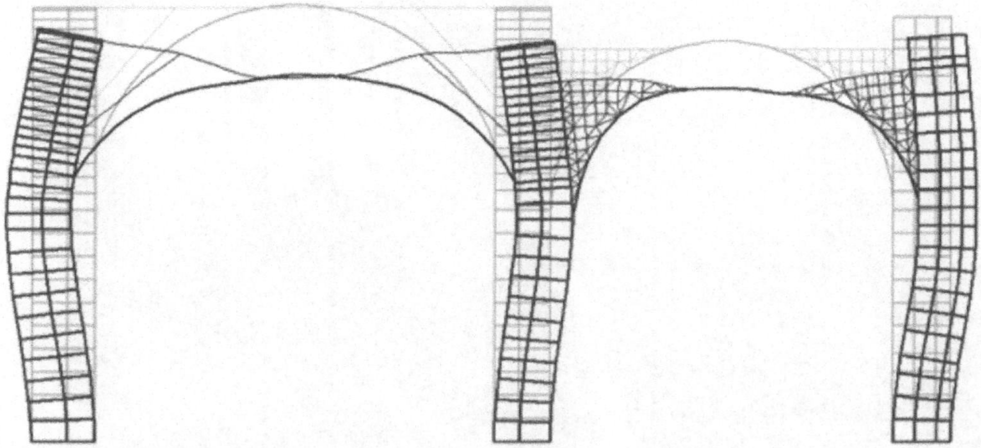

Figure 7.58 Static analysis: deformative states in the regime phase – Diagram by the author

of connections required for the pipes (Figures 7.59–7.60). From the perspective of conservation, the intervention could be reversible, while from the structural perspective, it makes it possible to avoid modifying the stress state of the construction plane of the foundations.

A further "improvement" has been made by eliminating the in-fills carried out by the military along the southern façade of the Citroniera in order to restore the link between the Orangery and the garden in front of it, allowing the light to flood in once more. The cantilevered balcony that damaged the imposing cornice was also eliminated, and work was carried out on the consolidation of the wooden trusses, as well as on the restoration of the roofing framework; the design of false ceiling "in the style of Juvarra" has enabled the existing trusses to be left intact.

Given the excellent state of the masonry fabric, plans were drawn up to carry out only local masonry repairs where necessary. Only in a few cases did it prove necessary to reconstruct the masonry; the materials and binders employed are traditional and completely compatible with the original ones. The improvement intervention on the vaults of the Citroniera and the Stables was designed to respect the original building and consisted of the following:

- Expose the surfaces of the extrados of the vaults; cleaning and stripping gaps;
- Cleaning and consolidating deteriorated parts;
- Configuration of a flat surface by refilling it with inert, light material;
- Execution of the base in reinforced concrete, secured to the extrados of the ribs.

The ceiling was significantly secured on the horizontal plane and bound, along the whole of the lateral perimeter, to the vertical external masonry (Figure 7.61).

To ensure access to many spaces of the old pavilion of the south-west façade, it was necessary to create a new floor above the old system of vaults. This floor, designed in reinforced concrete, was joined to the original masonry structure with a system of essentially reversible bonds (Figure 7.62).

Figure 7.59 Construction of the interior underground utilities – Photograph by the author

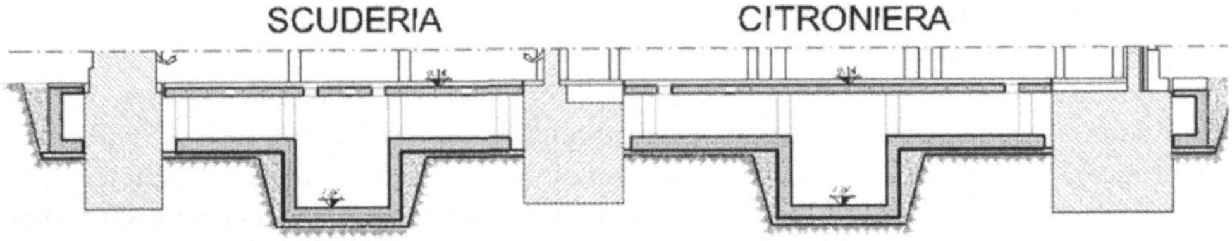

Figure 7.60 Cross-section of underground utilities – Drawing by the author

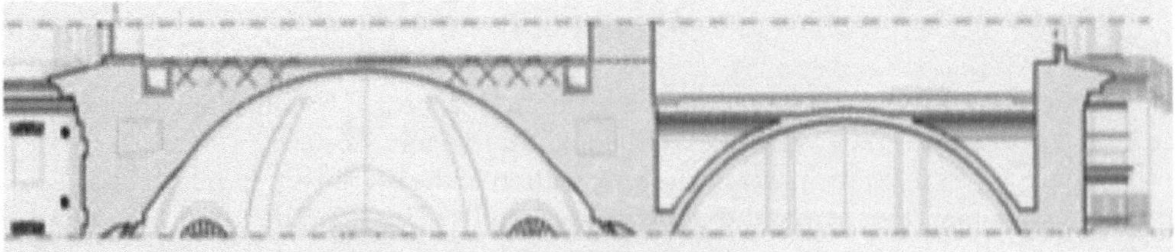

Figure 7.61 Cross-section of roof vaults – Drawing by the author

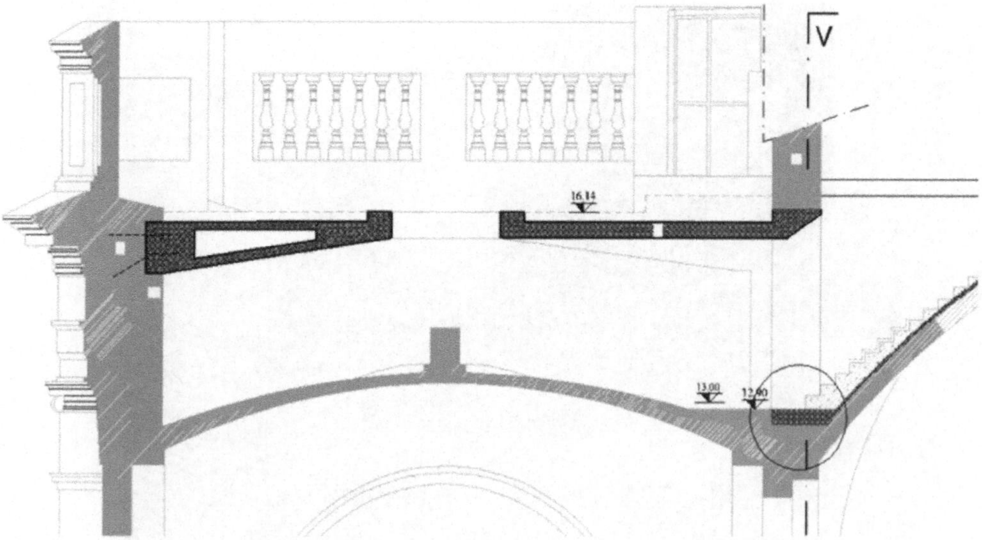

Figure 7.62 System of essentially reversible bonds joining the floor to the original masonry structure – Drawing by the author

In line with the above discussion, the improvement interventions were designed on those existing staircases that were in good condition, while it proved necessary to demolish one of the staircases, reconstruct it and create vertical connections with staircases and lifts. Once the improvement interventions had been designed and the overloads imposed by the new functions had been defined, the analysis of the finished elements of the three-dimensional model highlighted the excellent structural capacity of the historic monument to cope safely with its new use, fully respecting Juvarra's original design.

7.8.1.4 The implementation of the intervention

The execution of the structural project did not pose significant problems. Despite involving considerable amounts of excavation material, the construction of the underground utilities both inside and outside the complex took place without any adverse effects on the masonry; the same was true of the work to remove the infill walls on the southern façade of the Citroniera (Figures 7.63–7.64). The small-scale improvement interventions using reinforced concrete were always designed and carried out to ensure the maximum amount of reversibility and to avoid altering the original masonry fabric.

The conservative restoration of the Citroniera-Stables of the Savoy palace of the Venaria Reale succeeded in restoring the monumental complex to its original splendour, also creating new functions for the modern enhancement of the monument (Figure 7.65). This significant result was achieved at a structural level in the spirit of improvement, highlighting the excellent stability of the complex and largely respecting the original constructional design and techniques (Figure 7.66).

Figure 7.63 Construction of the outside underground utilities – Photograph by the author

Figure 7.64 Southern façade of the Citroniera after removal of the infill walls – Photograph by the author

Figure 7.65 The Citroniera after restoration – Photograph by the author

Figure 7.66 Façade of Citroniera after restoration (note the dormer windows) – Photograph by the author

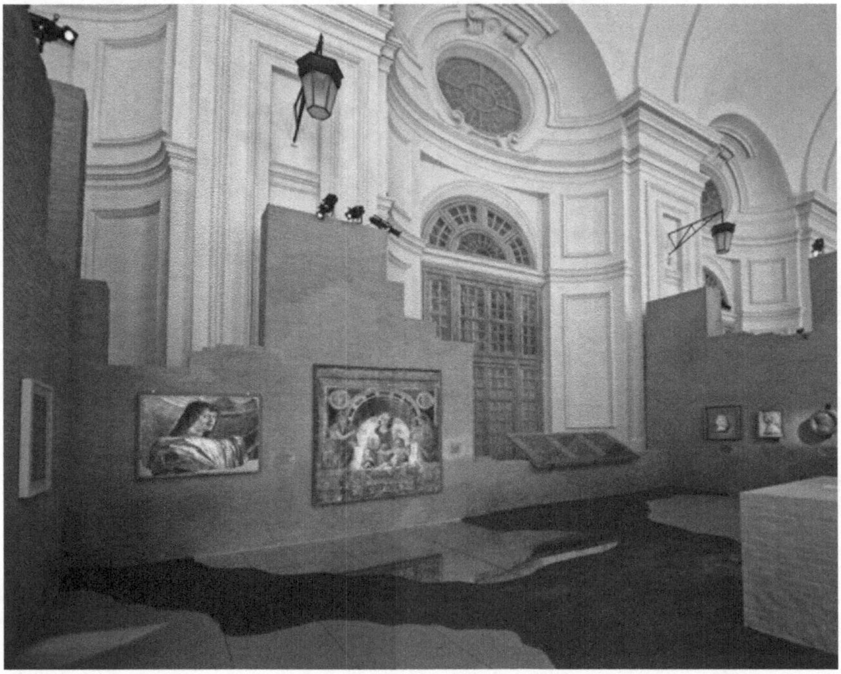

Figure 7.67 Exhibition held to celebrate the 150th anniversary of Italian unification – Photograph by the author

Its re-inclusion as a significant part of Italy's monumental heritage was formally recognised by a major exhibition held to celebrate the 150th anniversary of Italian unification (Figure 7.67).

7.9 An unusual example: The restoration of monuments on the Acropolis in Athens

Any overview of the last century of restoration in Europe should also illustrate the Greek experience, albeit succinctly. The advanced state of dilapidation, especially of the restored parts, of the majestic monuments of the Acropolis in Athens was the main reason for their general restoration which began over 30 years ago, giving new visibility to the glorious ruins, the custodians of Greek identity and the basis of Western civilisation. The restoration work culminated in the creation of a new archaeological museum at the base of the Acropolis.

The restoration work was carried out under the direction of the Committee for the Conservation of Acropolis Monuments (ESMA), which has carried out intensive work, presenting brief summaries of restoration projects for international debate. This scientific research has influenced the procedures for restoration for all the monuments of ancient Greece.

An enormous database has been created in which all the numerous stone structures on the level ground of the Acropolis have been identified, recorded and catalogued. This has

Figure 7.68 Athens, the Acropolis during restoration – *Tournikiotis: The Parthenon and its Impact in Modern Times*, Melissa. Athens 1994

enabled the numerous finds from the monuments to be carefully re-inserted during restoration work.

The site has gradually swollen in size to occupy an enormous area over recent decades, enabling large numbers of tourists to appreciate the various phases of this major undertaking (Figure 7.68). While respecting the theories of restoration and conservation as far as possible, the restoration work has followed a unique trajectory given the exceptional nature of the monuments built of Parian marble. Painstaking scientific research has been accompanied by lively discussion and debate within Greece and at an international level. Each decision has been carefully considered and evaluated during the execution stage. When the restoration work was completed, the huge complex was finally presented in all its splendour and majesty, ready to perpetuate its emblematic image through the centuries. Nevertheless, like every human endeavour, the work has led to doubts and even disagreement in the field. The following materials were used: Parian and Pentelic marble for additions and restoration; titanium, chosen to join the various elements, incorporating this innovative and corrosion-proof material into standard restoration procedure; natural artisanal stone for replacing sculptural elements; and lastly, white cement-based mortar with a low content of soluble salts. The choice of marble and titanium clearly appears to be an excellent one, while the use of natural artisanal stone and cement-based mortar may raise some question marks. The significant infilling of gaps was due to the need to ensure the stability of the reassembled monuments.

Three types of gaps were encountered:

a) Gaps caused by the elimination and the repositioning of elements mistakenly positioned in previous interventions;
b) Gaps formed due to the repositioning of scattered elements whose position has been identified;
c) Infilling of gaps in order to protect and maintain the parts of the existing structure.

Filling in gaps led to discussions about potential difficulties and risks that contrasted with restoration theory. However, in the case of the Acropolis, although they were significant and evident, the new additions did not affect the cultural value of the monuments, either historically or aesthetically (Figure 7.69) (Tournikiotis 1994).

The first restored monument was the Erecthion, which became the pilot monument for establishing the criteria and methods of restoration. In the Erecthion, 95 per cent of the blocks that had been previously repositioned proved to have been mistaken, while others were extraneous to the monument. Besides the material aspects, this reflects the complex work of scientific survey that was a feature of the entire restoration work. The inner surfaces of the blocks were seriously damaged due to fires in the past, so that filling in the gaps in order to make the new parts harmonise with the ancient ones proved to be an extremely complex process. The profile of the ancient piece was not modified but, with the aid of a pantograph, the face of each new element was shaped, partly bearing in mind

Figure 7.69 An example of new additions to a column of the Acropolis – *Tournikiotis: The Parthenon and its Impact in Modern Times*, Melissa. Athens 1994

Figure 7.70 An example of new additions to the temple of the Acropolis – *Tournikiotis: The Parthenon and its Impact in Modern Times*, Melissa. Athens 1994

the possibility of finding other ancient elements that would complete the monument. The in-fills were not treated from the perspective of natural ageing; over the decades, the "unseemly patching up" was gradually "absorbed" into the image of the ancient fabric, leaving the insertion of the new elements fairly evident. The decision not to apply an artificial patina will be a further element of distinction between the ancient and new parts which time is unlikely to cancel, while their perfect finish is due to the skill of the stone masons, who used the same tools as the original artisans. The criteria and choices applied to the Erecthion acted as a sample and an example for the Parthenon and for the other monuments of the Acropolis. Indeed, they tend to be used as a cultural reference point for all restoration work in Greece.

The work on the Pronaos of the Parthenon focused mainly on the legibility of the monument. Anaystilosis (the restoration of monuments using original elements and materials where possible) was carried out on six columns, of which three were complete and three were only partially preserved, up to a height of six or seven drums. The architrave of the constructed columns was repositioned, as was the architrave that connected the Pronaos to the southern wall of the cella of the temple. At the end of a long debate, the two capitals were finished "identically", while the new drums were not initially fluted; the fluting was subsequently undertaken only in correspondence to the ancient fluting on the original drums which is still legible. In this case, patination was undertaken to make the colour of the new marble as similar as possible to the ancient marble.

The restoration of the emblematic Temple of Athena Nike proved to be especially complicated. With Balanos' interventions, concrete was extensively used on the temple which, over time, had had disastrous consequences. The temple was completely dismantled down to the foundations and then reconstructed, eliminating all the concrete inserts, which were substituted by marble replacements. During the long and complex scientific and technical work, the Committee always emphasised that respect for the principles of minimum intervention, reversibility, easy recognition of the new elements and restoration according to traditional techniques and methods represented the crucial factor for contemporary interventions, which would be followed as far as possible.

The objective is not to alter the text and its potential importance as a document and source of evidence, but rather to guarantee physical and material consistency in order to ensure that the monument can be handed down to future generations. The gaps were therefore not filled in to conceal the perception of its image but to safeguard it and to facilitate its interpretation; it is a question of reducing damage without falling into the trap of falsification.

Notes

1 Archaeological Park of the Colosseum, the site of the church of S. Lorenzo in Miranda, fondo "Vittorio Ascoli Marchetti". Interventions directed by M.G. Filetici.
2 The restoration project of the Domus Tiberiana has been directed by M.G. Filetici since 2006.
3 Filetici M.G., Calabresi G., Ascoli Marchetti V., Carluccio G. (2007). *Il Bastione Farnesiano dopo il crollo del 1818–1829: costruzioni e interventi per la sua riconsegna alla città. In: Quale sicurezza per il patrimonio architettonico?* Mantua, Palazzo Ducale, 30 November–2 December 2006, Rome: Nuova Argos, pp. 408-427.ISBN/ISSN: 978-88-886-9312-5.

Filetici M.G. *Miglioramento statico o adeguamento? L'approccio storico e archeologico nel restauro della Domus Tiberiana al Palatino*. In: *Manutenzione e recupero nella città storica. Conservazione e normativa: esperienze recenti*, Atti del V Convegno Nazionale ARCo, Naples, Castello di Baia 11–12 giugno 2004, Rome: Gangemi, pp. 97–106. ISBN/ISSN: 88-492-0706-9

G. Calabresi, L. Cavalera, V. Ascoli Marchetti and M.G. Filetici, "Geotechnical aspects in the preservation of the Domus Tiberiana in Rome", in *Geotechnical Engineering for the Preservation of Monuments and Historic Sites*, Carlo Viggiani, ISBN: 978-1-138-00055-1, CRC Press, 2013, Print ISBN: 978-1-138-00055-1, eBook ISBN: 978-1-315-88749-4, DOI: 10.1201/b14895-24

Filetici M.G., TOMEI M. A. (eds.) *Domus Tiberiana, scavi e restauri 1990–2011*. Milan: Mondadori Electa. ISBN: 978-88-3708-935-1

S.A.G.-C.I.M. srl Roma carried out geodesic observations of the planimetric and altimetric movements of the structures overlooking via Sacra and monitoring of the movements related to the cracks (1985–2009: 2013-2016).

From the 1990s to 2009, Vittorio Ascoli Marchetti was part of the team working on the Domus Tiberiana and was responsible for the important interpretations of the conditions of the site and the interaction between the geological substratum, structural problems and geotechnical interpretations of the stability of the façade overlooking the Roman Forum and the Velabro valley. His last scientific contribution consists of the general report of 23.10.2009, SSBAR archive. Following his premature retirement from his scientific career, the "Fondo Ascoli Marchetti" was set up at the SSBAR archive containing the geotechnical data and documents related to the site until 2009; the material is the work of Ascoli Marchetti.

References

AA.VV., [1929], *Annales d'histoire economique et sociale*, now published as "Annales".
AA.VV., [1993], *Atti del Convegno "L'Athenaion di Paestum. Tra studio e restauro"*, Museo Archeologico Nazionale, Paestum.

AA.VV., [2005], *La Torre Restituita*, voll. I, II, III e Tavole, Bollettino d'Arte, Istituto Poligrafico dello Stato, Roma.

AA.VV., [2006], *La Torre Restituita*, vol. 4, volume speciale Bollettino d'Arte, Istituto Poligrafico dello Stato, Roma.

Adam J.P., [1983], *Degratation et restauration de l'architecture Pompeienne*, Edition du C.N.R.S, Paris.

Alberti B., Filetici M.G., Paris R., Santoro V., [2004], La tutela e il restauro del patrimonio archeologico nel parco dell'Appia Antica, in Centroni A., (ed.), *Conservazione e normativa: esperienze recenti*, Atti del V Convegno Nazionale ARCo (Castello di Baia, 4–5 giugno 2004), Rome, Gangemi, pp. 173–182.

Bellomo M., D'Agostino S., [1995], *The Contribution of a Technical-Scientific Culture in the Preservation of Archaeological Sites*, I Congresso Internazionale del CNR su "Scienza e Tecnòlogia per la Salvaguardia del Patrimonio Culturale nel Bacino del Mediterraneo", Catania.

Brigante M., [2018], Il tempio di Cerere, monitoraggio colonna 4, in De Palma G., (a cura di), *I templi di Paestum tra Restauro e Manutenzione*, Gangemi Ed., Rome.

Cairoli F.G., [2011], Provvedimenti antisismici nell'antichità, in *Journal Ancient Topography*, no. XXI.

Carbonara G., (ed.), [1981–1983], *Il cemento nel restauro*, Atti del Convegno AITEC, Roma, 1981.

Cipriani M., Avagliano G., [2007], *Il restauro dei templi di Poseidonia*, Atti del Convegno Internazionale, Paestum 26–27 giugno 2014, Ravenna, Valerio Maioli.

Como M., [2016], Sulla risposta sismica delle colonne, in D'Agostino S., (ed.), *Atti del VI Convegno nazionale di Storia dell'Ingegneria*, Naples, Cuzzolin.

D'Agostino, D'Ambrosio, Riccio, [2004]. 'Influenza degli interventi di consolidamento strutturale sul comportamento termoigrometrico del costruito antico' in F. Ribera (ed.) Luci tra le rocce. Colloqui internazionali "Castelli e citta fortificate". Storia, recupero, valorizzazione. Atti del Convegno. Alinea. Salerno.

D'Agostino S., [2014], Pompei, dal terremoto del 1980 ad oggi: costi e strategie, in *Economia della Cultura*, no. 3/4.

D'Agostino S., [2017], *Ingegneria per i Beni Culturali*, Bologna, Il Mulino.

D'Agostino S., Bellomo M., [1999], Excavation, Restoration and Conservation of Archaeological Sites. The Villa dei Quintili on the Appia Antica in Rome, in Jager W., Brebbia C.A., (eds), *Studies, Repairs and Maintenance of Historical Buildings VI*, Proceedings of STREMAH 99 International Conference, Southampton, WIT Press, pp. 451–460.

D'Agostino S., Cairoli Giuliani F., Conforto M.L., Guidoboni E., [2009], *Raccomandazioni per la redazione di progetti e l'esecuzione di interventi per la conservazione del costruito archeologico*, Naples, Cuzzolin.

D'Agostino S., Frunzio G., [1994], Structural Restoration of Archaeological Monuments in Seismic Areas, in Duma G., (ed.), *Proceedings of the 10th European Conference on Earthquake Engineering*, Rotterdam, Balkema, pp. 885–890.

D'Agostino S., Migliore M.R., [2006], *Il restauro costruttivo del complesso Citroniera-Scuderia a Venaria Reale: gli aspetti strutturali*, Atti del VI Convegno Nazionale ARCo "Quale sicurezza per il patrimonio architettonico?", Rome, Nuova Argos, pp. 587–595.

Dell'Orto Franchi L., (eds), [1990], *Restaurare Pompei*, Milan, Ed. Sugarco.

De Palma G., (ed.), [2018], *I templi di Paestum tra Restauro e Manutenzione*, Gangemi Ed., Roma.

Edwards P., [1777], *Capitoli che privatamente ni propongo alla considerazione degli EE.mirif.ri dello studio di Padova per l'effettuazione del Restauro generale Quadri di pubblica ragione, loro commesso con decreto dell'E.mo Senato 6 giugno 1771 (ante 10 agosto 1777).*

Fancelli P., [2004], Introduzione alla storia del consolidamento, in Rocchi P., (ed.), *Trattato sul consolidamento*, Mancuso, Roma.

Frontoni R., [2000], Appendice sulla tecnica edilizia, in MiBACT – Soprintendenza archeologica di Roma, *Via Appia. La Villa dei Quintili*, Rome, Electa, pp. 81–87.

Guidoboni E., Comastri A., Troina G., [1994], *Catalogue of Ancient Earthquakes in the Mediterranean Area up to the 10th Century*, ING – SGA, Bologna.

Guzzo P.G., [2001], *Pompei Scienza e società*, Milan, Electa.
Guzzo P.G., (ed.), [2003], *L'esperimento dell'autonomia, Pompei 1998–2003*, Milan, Electa.
Guzzo P.G., [2012], *Pompei. Appunti per una storia della conoscenza, coscienza e conservazione*, Naples, Bibliopolis.
Koldewey R., Puchstein O., [1899], *Die griechischen Tempel in Unteritalien und Sicilien*, Asher, Berlin.
Lizzi F., [1981], *Restauro statico dei monumenti*, SAGEP, Genoa.
Longobardi G., [2002], *Pompei sostenibile*, L'Erma di Bretschneider, Rome.
Masciari Genovese M., [1915], *Trattato di costruzioni antisismiche*, Hoepli, Milan.
Mastrodicasa S., [1948], *Dissesti statici delle strutture edilizie*, Milan, Hoepli.
Muñoz A., [1913], Via Appia Antica. Lavori di restauro e sistemazione del Ninfeo, in *BCAR*, XLI, pp. 14–24.
Paris R., (ed.), [2000], *Via Appia, la Villa dei Quintili*, Electa, Milano.
Paris R., [2001], "The Villa of the Quintili" in Archaeologia e Giubileo. gli interventi a Roma e nel Lazio nel Piano per il Grande Giubileo del 2000, vol. I, Naples. Electa.Rocchi P. (ed.), [2004], *Manuale del consolidamento*, Mancuso, Roma.
Sanpaolesi P., [1959], Il restauro delle strutture della cupola della Cattedrale di Pisa, in *Bollettino d'arte*, III serie, XLIV, pp. 199–230.
Sanpaolesi P., [1975], *Il duomo di Pisa e l'Architettura romanica toscana delle origini*, Nistri-Lischi, Pisa.
Severo D., [1997], *Filippo Juvarra*, Bologna, Zanichelli.
Tournikiotis P., (ed.), [1994], *The Parthenon and Its Impact in Modern Times*, Melissa Publishing House Atene.
Viscogliosi A., [2004], Impero romano, in Rocchi P., (ed.), *Trattato sul consolidamento*, Mancuso, Rome.
Von Riedesel J.H., [1771], *Reise durch Sizilien und Grogriechenland*, Orell, Geßner, Füßlin und Comp, Zurich.
Zuchtriegel G., (et al.), [2019], *Fragile Giants: The Monument of Paestum Between Safeguarding Knowledge and Fruition*, Conv. Int. di studi: monitoraggio e manutenzione delle Aree Archeologiche, Parco Archeologico del Colosseo, Rome.

Conclusions

The interest and attention focused on the conservation of monuments and built heritage have gradually grown in the last few decades throughout the world. United Nations, through UNESCO, has taken on the task of ensuring the protection of the main examples of universal material culture and its development, as is demonstrated by the increasing number of sites and monuments that have been declared World Heritage Sites.

However, as has emerged from a rapid overview of the situations in various countries, it is equally clear that the conservation criteria and the aims of interventions and policies that have been adopted differ, sometimes markedly, in each country due to the culture, history, physical, geological and climatic environment; the typology and historical age of the monuments; and the social needs linked to their use and function.

This variety of criteria and guidelines is also present in Europe in countries that have had a long shared historical and evolutionary development. It is clear, for example, that the conservation of archaeological sites and monuments poses specific problems that differ from those of Renaissance buildings and urban areas which are still frequented and inhabited, considering just safety, including seismic risk, and the growing demand for increasingly high levels of assurance.

These aspects of conservation increasingly tend to involve civil engineering, sometimes indirectly, due to the evolution and the gradual increase in the number of regulations which often clash with the design and construction criteria adopted in the past.

Geotechnical engineering, which is concerned with anchoring buildings firmly to the ground and their interaction with the surrounding environment, occupies a significant role in this field, partly due to the enormous strides made by this discipline in under a century in terms of our knowledge of soil mechanics and technological development. These advances have it possible to intervene in a wide variety of conditions and situations to increase the stability and safety of all buildings (the Leaning Tower of Pisa is an obvious example). In 1980, the International Society of Soil Mechanics and Geotechnical Engineering (ISMGE), at the behest of its former president, J. Kerisel, set up the Technical Committee for the Preservation of Monuments and Historic Sites in order to promote and provide an opportunity for the exchange of opinions between its members about various problems related to the conservation of important monuments and historic sites throughout the world. The committee has sponsored the publication of the proceedings of seminars, memoirs on specific issues and themes of general interest, and has launched the special Kerisel Lectures for international conferences held every four years.

Several significant differences in approach regarding the basic concepts and theories of conservation of built heritage have emerged among the engineers concerned during their

DOI: 10.1201/9781003160960-102

discussion of experiences in the geotechnical field. These differences have led to a greater focus on conservation theories and have prompted the publication of this volume.

The main issue regards the different sensibilities shown to each part of the work of architecture requiring conservation, including parts that are not visible, which constitute the material memory of the initial design and the modifications that have taken place over time, and which are of crucial importance. A purely iconic vision of the monument sometimes prevails that takes account solely of the visible parts; it is believed that substantial alteration, or even the replacement, of invisible structural components such as the foundations, despite being intrinsically linked to the design concept and executive techniques used at the time, do not conflict with the general concept of conservation.

An approach to conservation that shows more respect for the integral historical importance of monumental works of architecture is gradually becoming more widespread, as can be seen from the comparison of several major interventions carried out during the 1970s and more recent projects, but it is still not widely accepted among the organisations and institutions of various European countries which are responsible for safeguarding built and monumental heritage.

This brief overview reveals the complex historical process regarding the conservation of Europe's architectural heritage and the challenges posed by the need to pass it on to future generations.

The volume has examined the evolution of the concept of conservation and its innovative historical and scientific importance where the contribution of applied sciences, and, in particular, engineering in its various guises, is of essential importance.

A series of case studies displays the transition from a largely technical approach, which resulted in the reckless use of reinforced concrete, to the current approach of the monument-document of the historical archive of humanity, within a renewed vision of history as the science of humanity.

In particular, the examples show the possibility of proceeding by showing respect for the ancient concept of construction according to the principles of low-key minimal intervention and improvement which ensures the greatest possible degree of safety.

The volume then looked at the concept of programmed maintenance, which is the only effective tool for guaranteeing correct conservation and, as such, requires far more commitment from national governments.

Unfortunately, there are cases of inappropriate and invasive intervention such as the alleged anastylosis of a colonnade of the Temple of Peace in the Roman Forum.

The planned intervention "Safeguarding the Attic of the Colosseum" (*Per la messa in sicurezza dell'Attico del Colosseo*), which is also extremely invasive, will use reinforcing bars for the travertine blocks, prestressed wire strands on a glass bed and basalt FRP bars.

Until European universities train architects and engineers so that they are familiar with the hidden science that has allowed the construction of the glorious heritage built from remotest antiquity to the nineteenth century, arbitrary interventions will continue to be implemented, dominated by the improper use of innovative materials and heavily sponsored by the materials industry.

However, over the last few decades, as has been shown in the previous chapters and has just been scientifically proven (Como et al. 2019), this hidden science is now not just considerably better known but has also demonstrated its validity in the light of a correct reinterpretation according to the modern science of construction.

It is to be hoped that, on the one hand, the conservation of monumental heritage in a united Europe becomes a key commitment both in economic and cultural terms and, on the other hand, that effective integration takes place between a historical approach and a technical approach which, through the development of Archaeometry and Engineering for Cultural Heritage, collaborate fully towards our understanding and conservation of the historic built heritage of Europe. The challenges involved never cease. Major restoration work has just been completed on the Chapel of the Holy Shroud in Turin, a striking example of European Baroque architecture, and it has been carried out in a manner that displays the utmost respect for the authenticity of the monument. A complex cultural debate is also taking place that will involve the choices made for the restoration of Notre-Dame, an icon of European unity.

Bibliography

Como M., Iori T., Ottoni F., [2019], "Scientia abscondita." Arte e scienza del costruire Nelle architetture del passato, Marsilio, Venice, 2019.
Itaca F., "Il timido mestiere del conciatetti." Il Sole 24 Ore, Domenicale, 21 August 2017.

Bibliography

G. Calabresi and Salvatore D'Agostino

Abatino G., [1901], *La colonna del tempio di Hera Licinia in Capo Colonna* [Crotone], Id. MEFR XXIII, 1903, Naples.

Adam J.P., [1983], *Degradation et restauration de l'architecture Pompeienne*, Edition du C.N.R.S, Paris.

AGI (Associazione Geotecnica Italiana), [1991], *The Contribution of Geotechnical Engineering to the Preservation of Italian Historic Sites*, X ECSMFE, Florence.

Alberti B., (et al.), [2004], La tutela e il restauro del patrimonio archeologico nel parco dell'Appia Antica, in Centroni A., (ed.), *Conservazione e normativa: esperienze recenti*, Atti del V Convegno Nazionale ARCo, Gangemi Ed., Rome.

Alberti L.B., [1452], *De Re Aedificatoria*, Bollati Borlinghieri, Turin.

Arcoleo C., [1998], *Le ricette del restauro, malte, intonaci, stucchi dal XV al XIX secolo*, Marsilio, Venice.

Argan G.C., [1985], *Storia dell'Arte Italiana*, Sansoni, Florence.

Baratin L., (et al.), [2012], *Instruments and Methodologies for Cultural Heritage Conservation and Valorization*, Gabbiano Ed., Ancona.

Baratta A., [2007], Sulla statica delle scale in muratura alla romana, in *Notiziario Ordine degli Ingegneri della Provincia di Napoli*, no. 6.

Becchi A., Foce F., [2002], *Degli archi e delle volte*, Marsilio, Venice.

Bellomo M., D'Agostino S., [1995], *The Contribution of a Technical-Scientific Culture in the Preservation of Archaeological Sites*, I Congresso Internazionale del CNR su Scienza e Tecnologia per la Salvaguardia del Patrimonio Culturale nel Bacino del Mediterraneo, Catania.

Benvenuto E., [1981], *La Scienza delle Costruzioni e il suo sviluppo storico*, Edizione di Storia e Letteratura, Rome.

Bilotta E., (et al.), (ed.), [2013], *Geotechnics and Heritage*, Taylor & Francis Group, London.

Bloch M., [1949], *Apologie pour l'histoire ou métier d'historien*, A. Celin, Paris.

Borri A., Bussi L., [2011], *Archi e volte in zona sismica*, Editore Doppiavoce, Naples.

Boyd T.D., [1978], The Arch and the Vault in Greek Architecture, in *American Journal of Archeology*, vol. 82.

Brandi C., [2000], *Teoria del Restauro*, Einaudi.

Breymann G.A., [1885], *Il Trattato di Costruzioni Civili*, Vallardi, Milan.

Cairoli Giuliani F., [2006], *L'edilizia nell'antichità*, Carocci, Rome.

Cairoli Giuliani F., [2011], Provvedimenti antisismici nell'antichità, in *Journal Ancient Topography*, no. XXI.

Calabresi G., [2011], *The Soft Approach to Saving Monuments and Historic Sites*, Proceedings of European Conference ISSMGE, Athens, Balkema, Rotterdam.

Calabresi G., [2013], *The Role of Geotechnical Engineers in Saving Monuments and Historic Sites*, Rankine Lecture, Proceedings International Conference ISSMGE, Balkema, Paris.

Calabresi G., (et al.), [2013], *Geotechnical Aspects in the Preservation of the Domus Tiberiana in Rome*, Proceedings of the Second International Symposium on Geotechnical Engineering for the Preservation of Monuments and Historic Sites, A.A.Balkema.

Calabresi G., D'Agostino S., [1997], *Monuments and Historical Sites, Intervention Techniques*, Geotechnical Engineering for the Preservation of Monuments and Historic Sites, Balkema, Rotterdam.
Cangi G., [2005], *Manuale del Recupero Strutturale e Antisismico*, DEI Ed., Rome.
Cangi G., Caraboni M., De Maria A., [2010], *Analisi Strutturale per il Recupero Antisismico*, Ed. DEI, Roma.
Carbonara G., (ed.), [1981–84], *Restauro e cemento in Architettura*, AITEC, Rome.
Carbonara G., [1996], *Restauro architettonico*, UTET, Turin.
Casiello S., (ed.), [2008], *Verso una Storia del restauro. Dall'età classica al pieno ottocento*, Alinea Ed., Florence.
Caterina G., [1989a], *Tecnologia del recupero edilizio*, UTET, Turin.
Caterina G., [1989b], *Dissesti statici nelle strutture edilizie*, Hoepli, Milan.
Cavalieri di San Bartolo N., [1831], *Istituzioni di architettura statica e idraulica*, rist. nel 1855, Negretti, Mantua.
Cipriani M., Avagliano G., [2007], *Il restauro dei templi di Poseidonia*, Atti del Convegno Internazionale, Paestum 26–27 June 2014, Valerio Maioli, Ravenna.
Claudel J., Laroque L., [1870], *Pratique de l'art de construire*, Dunod, Paris.
Comitato Sismico del Ministero dei Beni Culturali, [1989], *Direttive per la redazione ed esecuzione di progetti di restauro comprendenti interventi di 'miglioramento' antisismico e 'manutenzione' nei complessi architettonici di valore storico-artistico in zona sismica.*
Como M., [2013], *Statica delle Costruzioni storiche in muratura*, Aracne, Rome.
Como M., [2016], Sulla risposta sismica delle colonne, in S. D'Agostino, (ed.), *Atti del VI Convegno Nazionale di Storia dell'Ingegneria*, Cuzzolin, Naples.
Como M., Iori I., Ottoni F., [2019], *Scientia abscondita*, Marsilio Editore, Venice.
Conforto M.L., D'Agostino S., [1995], Sulla concezione strutturale dell'architettura antica. Un caso emblematico: la Sostruzione romana, in *Atti del XII Congresso nazionale AIMETA*, tomo I, Giannini, Naples.
Conforto M.L., D'Agostino S., [1997], *Mechanical Resistance and Structural Behaviour of Some Constructive Prototype in the Mediterranean Area*, Proceedings of the 4th International Symposium on the Conservation of Monuments in the Mediterranean, Rhodes, 6–11 May, vol. 2.
Conti A., [1981], Conservazione, falso, restauro, in *Storia dell'Arte Italiana*, vol. 10, Einaudi, Turin.
D'Agostino S., [2003], L'Edificio storico: le strutture, in *Atti della Conferenza nazionale di Archeometria del Costruito*, Ravello, 6–7 February.
D'Agostino S., [2007], Il concetto di miglioramento e la sua evoluzione nella valutazione della sicurezza del patrimonio architettonico, in *Atti VI Convegno Nazionale ARCo, Quale sicurezza per il patrimonio architettonico?*, Mantova, Nuova Argos, Rome.
D'Agostino S., [2008], La storia del costruire e la meccanica razionale, in *Vv.Aa. Mathematical Physics Models and Engineering Sciences*, Liguori, Naples.
D'Agostino S., [2009], Sicurezza strutturale della colonna del Santuario di Hera Licinia a Crotone, in *Atti del VI Corso perfezionamento in Restauro Archeologico*, Alinea, Florence.
D'Agostino S., [2012], Terremoti e società: il ruolo dell'ingegneria, in *L'Acropoli*, anno XIII, no. 5.
D'Agostino S., [2014], Pompei dal terremoto dell'Irpinia del 1980 ad oggi: costi e strategie, in *Economia della Cultura*, Anno XXIV, nos. 3–4.
D'Agostino S., [2015], Between Mechanics and Architecture: The Quest for the Rules of the Art, in Aita D., Pedemonte O., Williams K., (eds.), *Masonry Structures: Between Mechanics and Architecture*, Birkauser-Springer International, Basel.
D'Agostino S., [2016], *La statica grafica e il calcolo delle strutture*, in D'Agostino S., (ed.), *Atti del VI Convegno Nazionale di Storia dell'Ingegneria*, Cuzzolin, Naples.
D'Agostino S., [2017], (ed.), *Ingegneria e Beni Culturali*, Il Mulino, Bologna.
D'Agostino S., Bellomo M., [1999], Excavation, Restoration and Conservation of Archaeological Sites. The Villa dei Quintili on the Appia Antica in Rome, in Jager W., Brebbia C.A., (eds.),

Studies, Repairs and Maintenance of Historical Buildings VI, Proceedings of STREMAH 99 International Conference, WIT Press, Southampton.

D'Agostino S., Frunzio G., [1994], Structural Restoration of Archaeological Monuments in Seismic Areas, in Duma G., (ed.), *Proceedings of the 10th European Conference on Earthquake Engineering*, Balkema, Rotterdam.

D'Agostino S., Giuliani C.F., Conforto M.L., Guidoboni E., [2009], *Raccomandazioni per la redazione di progetti e l'esecuzione di interventi per la conservazione del costruito archeologico*, Cuzzolin, Naples.

D'Agostino S., Guglielmo E., [1998], Un progetto di restauro tra conoscenza e conservazione: il Tempio di Diana a Baia, in *Atti del XIV Convegno Scienza e Beni Culturali*, Arcadia Ricerche, Marghera.

D'Agostino S., Marconi P., [1987], *Tecnologie di intervento nel restauro dei beni culturali*, Atti del I Seminario di Studi del Comitato Nazionale per la Prevenzione del patrimonio Culturale e la Questione Sismica, Il Ventaglio, Venice.

D'Agostino S., Migliore M.R., [2007], *Il restauro costruttivo del complesso Citroniera-Scuderia a Venaria Reale: gli aspetti strutturali*, Atti del VI Convegno Nazionale ARCo, Quale sicurezza per il patrimonio architettonico, Nuova Argos, Rome.

De Felice E., Sbrizioli E., Fedele R., [1990], *La Rocca dei Rettori di Bevevento, Rapporto tra storia e progetto*, Ed. Sintesi, Naples.

De Sivo B., [1992], *Il restauro degli edifici in muratura*, Ed. Dario Flaccovio, Palermo.

De Vecchi P., Vergani G.A., [2003], *La rappresentazione della città nella pittura italiana*, Ed. Silvana, Florence.

Dell'Orto Franchi L., (ed.), [1990], *Restaurare Pompei*, Ed. Sugarco, Milan.

Dinsmoor W.B., [1975], *The Architecture of Ancient Greece*, W.W. Norton & Company Inc., New York.

Di Pasquale S., [1996], *L'Arte del costruire tra conoscenza e scienza*, Marsilio Editore, Venice.

Di Teodoro F.P., [2003], *Raffaello, Baldassar Castiglione e la lettera a Leone X*, Minerva, Bologna.

Doglioni F., Moretti A., Petrini V., (eds.), [1996], *Le chiese e il terremoto*, Ed. LINT, Trieste.

Donghi, [1925], *Manuale dell'architetto*, UTET, Turin.

Erbani F., [2010], *Il disastro L'Aquila dopo il terremoto: le scelte e le colpe*, Laterza, Roma-Bari.

Ermentini M., [2007], *Restauro timido*, Nardini Editore, Florence.

Fancelli P., [2004], Introduzione alla storia del consolidamento, in Rocchi P., (ed.), *Trattato sul consolidamento*, Mancuso, Rome.

Ferretti M., [1981], Falsi e tradizione artistica, in *Storia dell'Arte Italiana*, vol. 10, Conservazione, falso, restauro, Einaudi, Turin.

Fiorani D., [2006], Trasformazioni del cantiere edile allo scorcio del Duecento, in *Franchetti Pardo, Arnolfo di Cambio e la sua epoca*, Viella, Rome.

Forest de Belidor B., [1729], *Le science des Ingenieurs*, A. Jomber, Paris.

Franchetti Pardo V., [1993], *Riflessioni e proposte sulla manutenzione dell'edilizia storica a scala urbana*, Atti del Convegno Manutenzione e recupero nelle città storiche, Rome.

Franchetti Pardo V., [2000], La memoria del passato prossimo, in *Il patrimonio architettonico del XX secolo fra documentazione e restauro*, Florence.

Franchetti Pardo V., [2002], Le discipline storiche nelle Facoltà di Architettura italiane: considerazioni e prospettive a valle delle esperienze fiorentine e romane, in Corsani G., Bini M., (eds.), *La Facoltà di Architettura di Firenze tra tradizione e cambiamenti*, Florence.

Franchetti Pardo V., [2003], Architettura, città, paesaggio, in *Convegno Nazionale di Studi Roberto Pane tra Storia e Restauro*, Naples.

Franchetti Pardo V., (ed.), [2006], *Arnolfo di Cambio e la sua epoca*, Viella, Rome.

Franchetti Pardo V., [2007], *Premessa e note su oltre centotrentanni di interventi edilizi alla SS.Trinità di Saccargia in SS. Trinità di Saccargia. Restauri 1891–1897*, Gizzi S., (ed.), Rome.

Franchetti Pardo V., [2014a], *Il Duomo di Orvieto analizzato iuxta sua principia, in La cattedrale di Orvieto: origini e divenire, Scritti editi e inediti*, Orvieto-Perugia.

Franchetti Pardo V., [2014b], Interrogativi sul Duomo di Orvieto, in *La cattedrale di Orvieto: origini e divenire*, Scritti editi e inediti, Orvieto-Perugia.

Franchetti Pardo V., [2014c], Nel settimo centenario dell'intervento del Maitani, in *La cattedrale di Orvieto: origini e divenire*, Scritti editi e inediti, Orvieto-Perugia.

Frontoni R., [2000], Appendice sulla tecnica edilizia, in *MiBACT – Soprintendenza archeologica di Roma, Via Appia*, La Villa dei Quintili, Electa, Rome.

Galasso G., [1998], *La Storia e le Storie*, Atti VII Convegno Italiano di Storia dell'Ingegneria, Cuzzolin, Naples.

Galasso G., [2000], *Nient'altro che storia*, Il Mulino, Bologna.

Galliazzo V., [1995], *I Ponti Romani*, Canova, Treviso.

Giavarini C., (ed.), [2005], *L'Erma di Bretschneider*, Rome.

Giovanetti F., (ed.), [1992], *Manuale del recupero di Città di Castello*, Ed. DEI, Roma.

Giovanetti F., Marconi P., (eds.), [1994], *Manuale del recupero della città di Palermo*, Flaccovio Editore, Palermo.

Giuffrè A., [1981], Pietà dei Monumenti, in Carbonara G., (ed.), *Restauro e Cemento*, AITEC, Rome.

Giuffrè A., [1988], *Monumenti e terremoti. Aspetti statici del restauro*, Multigrafica Editrice, Rome.

Giuffrè A., [1991], *Letture sulla meccanica delle murature storiche*, Ed. Kappa, Rome.

Giuffrè A., (ed.), [1993], *Sicurezza e conservazione nei centri storici*. Il caso Ortigia, Ed. Laterza, Bari-Rome.

Giuffrè A., Carocci C., [1999], *Codice di pratica per la sicurezza e la conservazione del centro storico di Palermo*, Ed. Laterza, Bari-Roma.

Greco G., (ed.), [1991], *Serra di Vaglio. La casa dei pithoi, Modena*, Franco Cosimo Panini, Modena.

Greco G., [2008], Costruire con la terra cruda: un esempio dell'antichità, in D'Agostino S., (ed.), *Atti del II Convegno Nazionale di Storia dell'Ingegneria*, Cuzzolin, Naples.

Greco G., [2010], La nascita del tetto con le tegole e la Formazione dei sistemi decorativi: qualche riflessione, in D'Agostino S., (ed.), *Atti del III Convegno Nazionale di Storia dell'Ingegneria*, Cuzzolin, Naples.

Gruppo nazionale per la difesa dai terremoti [GNDT], [1993], *Rilevamento della vulnerabilità sismica degli edifici in muratura*, Stampa tipografia Moderna, Bologna.

Guerrieri F., (ed.), [1999], *Manuale per la riabilitazione e la ricostruzione post-sismica degli edifici*, Ed. DEI, Rome.

Guidoboni E., [1989], *I terremoti prima del mille in Italia e nell'area Mediterranea*, SGA Ed., Bologna.

Guidoboni E., (et al.), [1994], *Catalogue of Ancient Earthquakes in the Mediterranean Area up to 10th Century*, ING-SGA, Bologna.

Guidoboni E., [2015], Contro la previsione. La radice culturale del primo progetto di casa antisismica, in *Quellen und Forschungen aus Italienischen Archiven und Bibliotheken 96/2016*, Deutschen Historischen Institut, Rome.

Guidoboni E., Bel J.E., [2009], Earthquakes and Tsunamis in the Past, in *A Guide to Techniques in Historical Seismology*, Cambridge University Press, London-New York.

Guidoboni E., Poriet J.P., [2019], *Storia culturale del terremoto dal mondo antico ad oggi*, Rubettino Saverio Mammelli, Catanzaro.

Guzzo P.G., [2001], *Pompei Scienza e società*, Electa, Milan.

Guzzo P.G., (ed.), [2003], *L'esperimento dell'autonomia, Pompei 1998–2003*, Electa, Milan.

Guzzo P.G., [2012], *Pompei. Appunti per una storia della conoscenza, coscienza e conservazione*, Bibliopolis, Naples.

Guzzoni D., (et al.), [2006], *Norme tecniche per le costruzioni*, Editore Il Sole 24 ore Norme & Tributi, Milan.

Hasckell F., [1981], La dispersione e la conservazione del patrimonio artistico, in *Storia dell'arte Italiana*, vol. 10, Einaudi, Turin.

Heyman J., [1996], *Arches, Vaults, Buttresses: Masonry Structures and Their Engineering*, Cambridge University Press, UK.

Heyman J., [1997], *The Stone Skeleton: Structural Engineering of Masonry Architecture*, Cambridge University Press, Cambridge, UK.

Kérisel J., [1987], *Down to Earth*, Balkema, Rootterdam, Boston.
Kérisel J., [2004], *Pierres et Hommes - Des Pharaons à nos jours*, Presses de l'Ècole National des Ponts et Chaussées, Paris.
Koldewey R., Puchstein O., [1899], *Die griechischen Tempel in Unteritalien und Sicilien*, Asher, Berlin.
Lamberti V., [1781], *La statica degli edifici*, Forgotten Books, Naples.
Lancellotta R., (et al.), (ed.), [2018], *Geotechnics and Heritage: Historic Towers*, Taylor & Francis Group, London.
Liberatore D., (ed.), [2000], *Vulnerabilità dei beni archeologici e degli oggetti esibiti nei musei*, CNR-Gruppo Nazionale per la Difesa dai Terremoti, Rome.
Lizzi F., [1981], *Restauro statico dei monumenti*, SAGEP, Genoa.
Longobardi G., [2002], *Pompei sostenibile*, L'Erma di Bretschneider, Rome.
Maltese C., [1960], *Storia dell'arte italiana 1785–1943*, Einaudi, Turin.
Marconi P., [1988], *Dal piccolo al grande restauro*, Marsilio Ed., Venice.
Marconi P., [1993], *Il restauro e l'architetto*, Marsilio Ed., Venice.
Marconi P., [1999], *Materia e significato. La questione del restauro architettonico*, Laterza, Roma-Bari.
Marconi P., [2012], *Restauro dei Monumenti*, Gangemi Editori, Rome.
Marino L., [2019], *Il restauro dei siti archeologici e manufatti edili allo stato di rudere*, Giannini e Figli Editori, Naples.
Martin R., [1980], *L'architettura Greca*, Electa, Milan.
Masciari Genovese F., [1915], *Trattato di costruzioni antisismiche preceduto da un corso di sismologia*, Hoepli Ed., Milan.
Mastrodicasa S., *Dissesti statici delle strutture edilizie*, Hoepli, Milan.
Melucco Vaccaro A., [1988], *Archeologia e Restauro*, Il Saggiatore, Milan.
Mertens D., [1984], *Aspetti dell'architettura a Crotone, in Crotone*, Atti del XXIII Convegno di Studi sulla Magna Grecia, Taranto.
Mertens D., [2016], Il tempio classico antico: concetti e progettazione, La ratio dei Greci nel costruire, in D'Agostino S., (ed.), *Atti del VI Convegno Nazionale di Storia dell'Ingegneria*, Cuzzolin, Naples.
MIBACT, [1988], *Progetto Pompei, primo stralcio, un bilancio*, Bibliopolis, Naples.
MIBACT, [1994], *Dopo la polvere. Rilevazione degli interventi di recupero [1985–1989] del patrimonio artistico monumentale danneggiato dal terremoto del 1980–81*, Tomo III, Istituto Poligrafico e Zecca dello Stato, Rome.
Milani G.B., [1923], *L'ossatura murale*, C. Crudo & C, Turin.
Milizia F., [1785], *Principi di architettura civile*, Forgotten Books, London.
Montanari T., [2013], L'età barocca. Le fonti per la storia dell'arte [1600–1750], Carrocci, Rome.
Muñoz A., [1913], Via Appia Antica. Lavori di restauro e sistemazione del Ninfeo, in *BCAR*, XLI.
Nola Molisi G.B., [1649], *Cronica dell'antichissima e nobilissima città di Crotone e della Magna Grecia*, Francesco Savio Stampatore, Naples.
Palladio A., [1570], *Quattro libri dell'architettura*, Ed. Mediterranea, Rome.
Paris R., (ed.), [2000], *Via Appia, la Villa dei Quintili*, Electa, Milan.
Pasta A., [1999], *Restauro conservativo ed antisismico*, Ed. Dario Flaccovio, Palermo.
Pergoli Campanelli A., [2015], *La nascita del Restauro*, Jaca Book, Milan.
Quatremère de Quincy, [1985], *Dizionario storico dell'Architettura*, Marsilio, Venice.
Regione Marche, AA.VV., [2004], *Vulnerabilità, manutenzione e progetto nel recupero post-sismico del patrimonio monumentale.*
Regione Marche, Doglioni F., (et al.),*Codice di pratica [linee guida] per la progettazione degli interventi di riparazione, miglioramento sismico e restauro dei beni architettonici danneggiati dal terremoto umbro-marchigiano del 1997*, published in B.U.R. Marche no.15 29/09/2000.
Regione Umbria, Vv.Aa., [2002], *Ricerche per la ricostruzione*, Ed. DEI, Rome.
Rocchi G., [2002], *La Basilica di San Francesco ad Assisi prima, durante e dopo il terremoto del 1997*, Alinea, Florence.
Rocchi P., (ed.), [2004], *Manuale del consolidamento*, Mancuso, Rome.

Rondelet J.B., [1802], *Traité Théorique et pratique de l'art de batir*, Edizione italiana del 1831, Nabu Press, Florence.

Ruskin J., [1987], *Le pietre di Venezia*, Rizzoli, Milan.

Sanpaolesi P., [1959], Il restauro delle strutture della cupola della Cattedrale di Pisa, in *Bollettino d'arte*, III serie, XLIV.

Sanpaolesi P., [1975], *Il duomo di Pisa e l'Architettura romanica toscana delle origini*, Nistri-Lischi, Pisa.

Scamozzi V., [1615], *L'idea dell'architettura universale*.

Severo D., [1997], *Filippo Juvarra*, Zanichelli, Bologna.

Strazzullo F., [1991], *Restauri del Duomo di Napoli tra '400 e '800*, Fondazione Corsicato, Naples.

Studio sulla *vulnerabilità sismica e proposta di interventi per un centro storico attraverso l'indagine tipologica e l'utilizzo di un database georeferenziato*, Ricerca condotta dal Dipartimento di Ingegneria Civile ed Ambientale dell'Università degli Studi di Perugia con la collaborazione del Comune di Città di Castello, 2002.

Todaro G., [2009–2010], *Attività di manutenzione e cura nei Beni Culturali-Architettonici*, Degree dissertation, Politecnico di Milano.

Toraldo F., Chianese A., Rosati L., [2016], Abaco di progetto di parete muraria a gravità nell'area archeologica di Pompei, in D'Agostino S., (ed.), *Atti del VI Convegno Nazionale di Storia dell'Ingegneria*, Cuzzolin, Naples.

Tournikiotis P., (ed.), [1994], *The Parthenon and Its Impact in Modern Times*, Melissa Publ. House, Athens.

Urbani C., [2000], *Intorno al restauro*, Skyra, Milan.

Valadier G., [1832], L'architettura pratica dettata nella Scuola e Cattedra dell'Insigne Accademia di San Luca.

Viggiani C., (ed.), [1997], *Geotechnical Engineering for the Preservation of Monuments and Historic Sites*, Balkema, Rotterdam.

Viggiani C., [2019], *Senza neanche toccarla – La stabilizzazione della Torre di Pisa*, Heveluius Ed., Benevento.

Viollet-le Duc E., [1854–1868], Restauration, in *Dictionaire raisonné de l'architecture francaise du XIe au XVIe siecle*, vol. VIII, Bance-Morel, Paris.

Viollet-Le-Duc E., [1982], *L'architettura ragionata: estratti dal Dizionario di E. Viollet-le-Duc*, Jaca Book, Milan.

Viscogliosi A., [2004], Impero romano, in Rocchi P., (ed.), *Trattato sul consolidamento*, Mancuso, Rome.

Vitruvio M. Pollione, [1997], *De Architectura*. Many cheap editions exist [ensure that they contain all 10 books].

Von Riedesel J.H., [1771], *Reise durch Sizilien und Grogriechenland*, Orell, Geßner, Füßlin und Comp, Zurich.

Vv.Aa., [2006], *La Torre Restituita*, vol. 4, Volume speciale Bollettino dell'Arte, Istituto poligrafico dello Stato, Rome.

Zingone A., [2012], S. Leucio da utopia preindustriale a polo culturale. Un restauro esemplare, in D'Agostino S., (ed.), *Atti del IV Convegno Nazionale di Storia dell'Ingegneria*, Cuzzolin, Naples.

Index

Note: Page numbers in *italics* indicate figures.